THE BEST
OF THE
HOLLOW HASSLE

VOL. 2 - #4

Malice in wonderland

VOL. 3 - NO. 2

VOL. 5-NO.3   MAY 1, 1984

GLOBAL COMMUNICATIONS/CONSPIRACY JOURNAL

THE BEST OF THE HOLLOW HASSLE

Timothy Green Beckley: Editorial Director
Carol Rodriguez: Publishers Assistant
Sean Casteel: Associate Editor
William Kern: Editorial Assistant
Author: Mary J. Martin & Tim R. Swartz
Cover Art: Tim R. Swartz

Printed in the United States of America

For free catalog write:
Global Communications
P.O. Box 753
New Brunswick, NJ 08903

Free Subscription to Conspiracy Journal E-Mail Newsletter
www.conspiracyjournal.com

THE BEST OF THE HOLLOW HASSLE

CONTENTS

THE HOLLOW HASSLE

VOL. 3 - NO. 1   OCT. 31, 1981

CREDIT: MANLY PALMER HALL

Take Me Down To The World Beneath Our Feet
By Tim R. Swartz

As theories go, the idea that the Earth is hollow does not garner a lot of respect. For most people, the hollow Earth is probably a close second to the "Earth is flat" theory on that big list of "crack-pot" ideas. Nevertheless, as long as there have been people able to sit around a camp-fire, tales of a mysterious inner world have been part of mankind's heritage.

What is it about the hollow Earth theory that continues to fascinate people? Perhaps it is because people love a good mystery and right now there are not a lot of good mysteries left for people to cling to. The surface of the planet has been almost completely explored, and now we are taking those first steps to penetrate the vast reaches of outer space. So what does that leave for the rest of us who have that primal urge to see what lies on the other side of the mountain?

The deepest parts of the oceans are still almost completely untouched by human exploration, but it is not so easy for most of us to do that sort of exploring. You either have to have a whole lot of money or the ability to hold your breath for a really long time to do any serious undersea exploration. I think that is why I have always been so fascinated with the whole idea of the Hollow Earth, especially the idea that you could access it via long-lost secret tunnels that connected to the many caves and caverns that criss-cross the planet just underneath the surface.

One part of the Hollow Earth theory that was especially interesting to me was the "Shaver Mystery," first introduced to me by Timothy Green Beckley's book *The Shaver Mystery and the Inner Earth*, originally published by Gray Barker's Saucerian Press. With the Shaver Mystery you had an ancient race of mutated, crazed, underground dwelling humanoids whose sole purpose in their twisted lives was to make the existence of those who dwell on the surface a living hell.

Where a majority of books about UFOs and their extraterrestrial pilots were all peace, love, spiritual growth and harmony; Richard Shavers stories were the polar opposite. The dero didn't want peace and harmony - they wanted death and destruction. The Venusians were here to save us from our own atomic weapons - the dero wanted the choicest and juiciest of our population for lunch. The Martians wanted to teach us new philosophies and personal growth - the dero used their ray machines to crash our planes and steal candy from our babies.

Yep, the dero were some pretty badass underground mutants. And I ate it all up with a spoon and asked for more.

At this point in my young life, I was already familiar with the whole hollow Earth theory thanks to writers such as Brad Steiger and Raymond Bernard; but once again, as with the flying saucer mythology, the hollow Earth was allegedly filled with utopian societies that had long ago abandoned such crass things as money, war, meat-eating, and cold beers on hot summer nights. The dero, on the other hand, sounded like guys that would have enjoyed a keg of beer while they were torturing their large-breasted human slave women.

This is why I am happy to introduce to you the best of Mary J. Martin's *The Hollow Hassle*, a "zine" that during its day dealt almost exclusively with the hollow Earth mystery, especially Shaver's ideas and the prospect that there were secret caverns out there that could lead to the secret underground cities. Most of the articles included in this book have not been seen since they were originally published and they offer a plethora of valuable information that anyone interested in the hollow Earth and its related theories will find extremely fascinating.

But be warned...those who have delved too deeply into the mysteries of the inner Earth have often paid a heavy price for their curiosity. With Richard Shaver and his deros, it is the darkest corners of mankind's psyche that rules supreme in the secret tunnels deep underground. So take care when you go exploring Richard Shavers inner world, you never know what terrors and delights wait just beyond the corner.

VOL. 5 - NO. 4

AUG. 1, 1984

CHAPTER ONE
The Hollow Earth Hassle
Of Mary Jane Martin
By Sean Casteel

For "technical purposes," Mary Jane Martin is the editor of the work you are holding in your hands. The material in this publication is composed of letters and published articles that were used in the *Hollow Hassle* newsletter that Mary Jane put out for several years. She knew everyone in the Shaver-Hollow Earth field, a field that was small and limited due to its frankly bizarre nature.

Mary is the last of the Richard Shaver "inner circle." We felt it important that we work with her in putting this material together into a practical framework so that it might be available to other students of these remarkable subjects.

Of the original participants in the drama that came to be known as the Shaver Mysteries, very few remain alive today. Both Richard Shaver and his mentor, Raymond Palmer, passed away in 1975, each of them steadfastly maintaining that the work they had published about the Tero and Dero of the Hollow Earth was factual, though its heyday with diehard believers had long since passed.

Among the handful of believers who have kept the subject of the Shaver Mysteries alive and "out there," it was left to Mary Jane to begin publishing a newsletter in the early 1970s devoted to the latest findings about the dark kingdom below our feet that sends such endless misery to us hapless mortals on the surface.

In an interview conducted for this book, we spoke to Mary Jane and asked how she first came to the subject. She said it all started when she saw an ad for *The Smoky God*, Or *A Voyage To The Inner World*, a 1908 novel by Willis George Emerson which is presented as a true account of a Norwegian sailor who sailed through an entrance to the Earth's interior at the North Pole.

"I can't recall what the ad said," Mary Jane said, "or what piqued my interest enough to send for it, but I have never regretted that I did. This was my first encounter with the theory of a Hollow Earth. I was also receiving a newsletter put out by Dorothy Starr, with its main interest being in the possibility of a polar shift. I wrote to Dorothy and asked her if any of her readers were interested in the theory of a Hollow Earth. She had one member who shared my interest, namely Tom LeVesque, who I started to correspond with."

It was around that time that Mary Jane began printing her own small newsletter called *The Hollow Hassle*. LeVesque became her coeditor, and the newsletter was published monthly for a total of thirteen issues. Mary Jane and LeVesque were married in 1973, and began to travel around the country investigating paranormal phenomena, including cattle mutilations.

Mary Jane said that in the wake of reading *The Smoky God* and another book called *Etidorhpa* (an 1898 novel by John Uri Lloyd about a Freemason who descends to the Inner Earth through a cave in Kentucky and encounters strange forms of life there), she knew that solid, real-world information to back up the claims of Shaver and others would be hard to come by.

"I knew that it would be a 'hassle' to find any information on this," she explained, "thus the title '*The Hollow Hassle*.'"

Due to their travel schedule and dwindling funds, publication of the newsletter ceased for a time. After she and LeVesque separated in 1981, Mary Jane began to print the newsletter again as a quarterly, running for about another seventeen issues into the mid-1980s.

It was a gut feeling about Richard Shaver's writings that kept her believing for so long.

"There was a lot of factual evidence to show that a Hollow Earth had a good possibility of being true," she said. "Many people are more versed in the scientific aspects than I am, but you do have things like the Narwhales that travel north and disappear. Many legends talk of their ancestors disappearing further north, possibly entering through the North Pole entrance, just as the Narwhales probably are."

Along the way, Mary Jane struck up a relationship with Shaver himself.

"Initially," she said, "I started to correspond [by regular mail] with Dick Shaver, but I did go to his Arkansas home twice. The first time, he looked healthy, but the second visit Dick looked thinner and worn down. I also did a radio talk show interview with Dick, on the Hilly Rose Show. Tom and I did the first show by ourselves, and on the next show, Dick came in at about the halfway mark."

Mary Jane also came to know others who were likewise in pursuit of the truth about the Shaver Mysteries.

"I was very fortunate to be in touch with a number of very interesting people," she said. "They included Richard Toronto, of Shavertron fame, Charles Marcoux, who died on the way to Blowing Cave in Arkansas, Joan O'Connell from *The New Atlantean Journal*, and Bruce Walton, now known as 'Branton,' who was an excellent researcher on the Hollow Earth/Inner Earth."

Did Mary Jane ever do any exploring of caves herself?

"I have explored some caves," she replied, "and I also went to a number of tourist caves. Unfortunately, I did not find a tunnel that led to the Inner Earth, but many of the tourist caves have areas where you are not allowed to go, so who knows how far those areas go? I do feel that the Mammoth Cave in Kentucky may be an entrance."

That there is something wicked lurking in the hollow places is accepted without question by Mary Jane.

"I do believe there is evil in those caves," she said, "and Richard Shaver warned us not to go into the caves. There are too many crazy things that happen on Earth, with no apparent reason for them to happen. One man got hit by lightning twice in his life. What are the odds of that? Young people do things that are very evil and they seem to have no control over their actions. Deros would explain a lot of things. Also, when Ray Palmer stayed overnight at Richard's house, he heard voices."

Still, a certain amount of controversy continues to enshroud the relationship of Shaver and Palmer and the literal truth as understood between them. For example, Palmer once declared that during the time Shaver was allegedly underground, Shaver was instead in a mental institution in a coma. We asked Mary Jane for her thoughts on the controversy.

"That's a tough question," she said. "A lot of people think it was an astral trip that Richard took while being in a coma, and that may be. But whether it's an actual trip or an astral one, on some level it is real. It certainly was to Richard, and I'm inclined to believe him. We got pretty close in the years that we knew each other.

"I feel that Richard wrote what he believed to be true," Mary Jane continued, "and then Palmer dressed it up to make it more exciting and saleable. Palmer only paid Richard one and a half cents per word, and I think there was trouble about that."

Much of the unhappy events that have transpired in the years since Mary Jane blames quite firmly on the Dero, including her own health problems, among other things.

"At one time," she said, "I do think they were stopping me from putting out *The Hollow Hassle*, as I had been rushed to the hospital twice with chest pains. It wasn't just me that was being hit by the Deros. Joan O'Connell, who had published more Hollow Earth info, died from a heart attack. She had no heart problems previously. Then her material went to Gray Barker, who was going to continue publishing her newsletter. He died three months after Joan, from a heart attack, I believe. Then Charles Marcoux was killed by a number of bees as he was heading towards Blowing Cave in Arkansas."

But nothing tops the death of Richard Shaver himself when it comes to strange factors being at play.

"Richard Shaver's death was highly suspect also," Mary Jane said. "He had signed a contract to come to Hollywood and be a consultant on a movie about the Teros and Deros. He was very excited and even got new dentures for the occasion. Then he came down with locked bowels and had to go into the hospital. He told his wife Dot that the Deros would not let him make this movie and he would die in the hospital. She assured him that it was not a serious operation and that he would be fine."

Dot was quite correct that the surgery itself would go well. But as he lay recovering in his hospital bed, Shaver suffered a pair of small heart attacks.

"Neither of which should have killed him," Mary Jane said, "but he died in that hospital. His prediction came true. The Deros would not let him live to make that movie. Afterwards, Dot could not find the contract for the movie or who was making it, as we wanted to check into that further."

Finally, Mary Jane said the Hollow Earth may have been created initially as the Garden of Eden, despite the scoffing of modern day science.

"Science is wrong a lot," she said, "and they still don't really know much about what is below our feet. Someone told me that if God was going to build a house for Adam and Eve, would he build it so they had to live on the porch outside the house, or would he build it where there was no glaring sun, with large gardens of fruit, stable temperatures, etc., such as the Hollow Earth? Perhaps when Adam and Eve got thrown out of the Garden, it meant that they were tossed out of the Inner World, to suffer the extreme climates on the outside of the Earth."

Were we indeed cast out of an Inner Earth paradise when we fell from grace? Like so much of the Shaver Mysteries to which Mary Jane has devoted years of study, the answer may lie in some hidden place that only Richard Shaver and a few others have ever reached. In any case, we are left to struggle against the Dero ourselves, or whatever the embodiment of evil that still tortures mankind is ultimately called. And Mary Jane's place in that overall scheme of things can be read right here, in the pages of yet another attempt to make sense of the unknowable before it is too late to make sense of anything.

OCT. 31, 1980

"My name is Etidorhpa. In me you behold the spirit that elevates man, and subdues the most violent of passions. In history, so far back in the dim ages as to be known now as legendary mythology, have I ruled and blessed the world. Unclasp my power over man and beast, and while heaven dissolves, the charms of Paradise will perish. I know no master. The universe bows to my authority. Stars and suns enamored pulsate and throb in space and kiss each other in waves of light; atoms cold embrace and cling together; structures inanimate affiliate with and attract inanimate structures; bodies dead to other noble passions are not dead to love. The savage beast, under my enchantment, creeps to her lair, and gently purrs over her offspring; even man becomes less violent, and sheathes his weapon and smothers his hatred as I soothe his passions beside the loved ones in the privacy of his home.

"I have been known under many titles, and have comforted many peoples. Strike my name from Time's record, and the lovely daughters of Zeus and Dione would disappear; and with them would vanish the grace and beauty of woman; the sweet conception of the Froth Child of the Cyprus Sea would be lost; Venus, the Goddess of Love, would have no place in song, and Love herself, the holiest conception of the poet, man's superlative conception of Heaven's most precious charms, would be buried with the myrtle and the rose. My name is Etidorhpa; interpret it rightly, and you have what has been to humanity the essence of love, the mother of all that ennobles. He who loves a wife worships me; she, who in turn makes a home happy, is typical of me. I am Etidorhpa, the beginning and the end of earth. Behold in me the antithesis of envy, the opposite of malice, the enemy of sorrow, the mistress of life, the queen of immortal bliss.

VOL. 2 - NO. 1

CHAPTER TWO
Follow Me To The Center Of The Earth
VOL. 2 NO. 1 – OCT. 1980

Eccentric John Symmes was all set for his epic journey to the Earth's core. All he needed was "100 brave men" as companions-and unlimited funds from the United States government.

Eccentricity and inventiveness seem to be elements of the Anglo-Saxon mentality and it is not surprising therefore that men of English stock who emigrated to America took their inventive eccentricities with them.

One of the earliest attempts to fuse scientific ideas with theology was made by Cotton Mather, the son of a 17th-century English clergyman, who was better known for his passionate pursuit of witches in Salem, Massachusetts, but who was among the first men to suggest that the Earth might be hollow.

All manner of curious ideas have been advanced by original thinkers relating to the mysteries of the Earth's interior. A Jesuit named Hardouin once advocated the ingenious theory that Hell lay at the center of the Earth and that the rotation of the globe was the result of hordes of damned souls everlastingly attempting to clamber up its sides to escape the flames, their combined weight causing the world to rotate "just as a squirrel turns his cage or a dog turns the spit."

The Hollow Earth theory seems to have inspired undying interest for centuries. The renowned astronomer Edmund Halley had his own views on the subject, contending that the Earth possessed a shell 500 miles thick protecting an interior which was capable of sustaining life. However, it was not until the 19th century, the great age of science, that the idea of a hollow Earth began to attract a wide public, and it was then due to the efforts of an American eccentric named John Clare Symmes who set in motion a chain of speculation which still continues today.

Symmes was no scientist but a soldier with a fine record in the American forces. He had served as an officer in the war of 1812, the last in which Britain and America fought against each other, and had afterward retired from the army to give all his attention to the mysteries of the Earth's interior.

Crystal Clear

From his researches, Symmes gradually developed the theory that the world was not only spherical but hollow. It consisted of five concentric shells each set within the other and there were openings several thousand miles in diameter situated at each of the North and South Poles. The logic of the idea appeared so crystal clear to Symmes that he seriously put forward the proposition that an expedition should be sent to investigate the mysterious 'holes' at first hand. He issued a rousing call for support which he trusted would lead to an immediate response from the more adventurous members of the American population, demanding '100 brave companions' with himself as their leader.

Symmes launched his appeal in 1818 and breathlessly awaited the outcome. But it produced the loudest horse-laugh in American history. Angrily, Symmes argued that the expedition would confirm beyond every shadow of doubt the existence of plant and animal life in the Earth's interior "as well as on the convex surface of the next sphere." His confidence in this theory was absolute. With what has been described as "the assurance of a certified visionary" he set out to raise funds for the great adventure, traveling the length and breadth of the United States exhorting the public to support his dream. He actually petitioned the United States government for the necessary funds, and strange to say not only were two petitions tabled by Congress but the second received 25 votes in its favor.

Nasal Voice

However, the strain of stomping up and down the country for many years addressing public meetings with the fervor of a Messiah was gradually wearing Symmes down. For one thing he lacked the personality for a successful campaign. He was certainly no orator, having a "stumbling nasal voice." The struggle gradually proved too much for him and his health collapsed. Symmes died at

Hamilton, Ohio, in 1829. His son Americus Symmes, one of his father's staunchest supporters, dutifully set up a memorial over the grave surmounted by a stone model of Symmes's hollow Earth, and then took over the campaign for recognition of his father's theories.

Americus Symmes proved to be a powerful upholder of the Hollow Earth theory and published a book, The Symmes's Theory of Concentric Spheres, in 1878 in which he assembled a series of powerful arguments in its favor, drawing data from "physics, astronomy, climatology, the migratory habits of animals and the reports of travelers." Americus summed up his entire case with the punch line: "Reason, common sense, and all the analogies in the natural universe conspire to support the theory."

Americus Symmes managed to enlist a handful of converts who seemed to have regarded the subject in what can only be described as a whimsical manner. Apparently, some of them imagined the hollow Earth to be a gigantic receptacle within which the Creator stowed away all the natural objects for which there was no immediate use at the time. As one of them gracefully put it: "A hollow Earth habitable from within would result in a great saving of stuff." The first notable reaction to Symmes's theory belongs to the area of ironic science fiction rather than pure science. In 1820 a book was published under the pseudonym 'Captain Seaborn', with the title *Simozio*. It told the story of a ship and its passengers who, venturing into the region of the South Pole, were sucked by a powerful current into one of Symmss's Holes where it glided gracefully into the center of the Earth. Here everyone aboard discovered, much as Symmes had predicted, a rich and varied form of life, in fact it was a socialistic Utopia with a highly civilized population who dressed in white and practiced the belief of the brotherhood of Man. Others then seized on the profitable theme which has attracted science fiction writers ever since. There is good reason for believing that Jules Verne's *Journey to the Center of the Earth* was to some extent influenced by the Symmes's syndrome.

Symmes's body might have moldered in his grave in Ohio but his theory went soldiering on into the following century, collecting a miscellany of eccentric, scientific and religious supporters.

In 1913, about a century after Symmes had failed to take the world by storm, Marshall B. Gardner of Illinois, a maintenance engineer in a corset company, took up the banner with a personal interpretation of the theory which he published in his book *Journey into the Earth's Interior*. While rejecting the idea of concentric spheres, he argued that the hollow Earth was protected by an outer shell 800 miles thick, which contained a sun some 600 miles in diameter. He agreed with Symmes as to the existence of vast openings at the Poles but contended that these were blocked by huge ice cape. Among the original inhabitants of the hollow Earth were Eskimos and mammoths whose frozen bodies had been discovered in Siberia. Utterly committed to his theory, Gardner published an expanded edition of the book in 1920, new 456 pages in length, in which he asserted that the Earth was illuminated by an internal sun which had been mistaken for the Aurora Borealis. Inevitably Gardner was written off by the scientists of his day as another Symmes. He was plainly hurt by their lack of vision, which seemed to baffle him, for he wrote plaintively: "If you are not one of them they do not want to listen to you."

Surprises in Store

It is amusing to consider that Gardner regarded his predecessor Symmes as little mere than a "scientific pretender" and a crank and that he remained unconvinced of any errors on his own part even after Admiral Richard Byrd had flown over the Pole and failed to discover any of the expected 'Symmes's Holes' in the Earth's surface. Gardner died in 1937 still adhering to the remnants of a theory which was now tending to bore the public. However, there were other surprises in store.

From the year 1870 onward for the next 38 years an American mystic named Cyrus Teed wandered the length and breadth of his country presenting his own version of the Symmes's Hollow Earth theory in the form of a new religion.

Cyrus Teed claimed to have received a beatific vision in which a mysterious woman had suddenly materialized in his laboratory and imparted the information that not only was the Earth hollow but that the entire human race lived inside it instead of on the surface as had been hitherto supposed. Furthermore, Teed had been commissioned by the divine powers to make known the following essential truth to the entire world: "All existence is inside the Earth including the sun, moon, stars and comets. Everything is inside; nothing at all is outside," to quote Martin Gardner's Fads and Fallacies in the Name of Science. Teed insisted that the Earth's crust was 100 miles thick and comprised 17 separate layers.

Unlike the mumbling Symmes, Teed had a flair for florid oratory which captivated the huge audiences who flocked to hear his revelations. His inspired followers had to accustom themselves to a flood of esoteric verbiage interlaced with 'scientific' terminology especially when he began to discuss comets as the result of "crusic force caused by condensation of substance through the dissipation of the colorific substance at the opening of the electromagnetic circuits which closes the conduits of lunar and solar energy."

Stressing the religious aspect of his theory, Teed thundered: "To know of the Earth's concavity is to know God. To believe in the Earth's convexity is to deny Him and all His works." Cyrus Teed called his new religion 'Koreshanity' after the name he had adopted for himself – Koresh being the Hebrew word for Cyrus. All who refused to accept the dynamic doctrine of the Hollow Earth, and particularly those who actively opposed the idea, were accused of being limbs of Antichrist.

In his role of Messiah, Teed founded a College of Life in Chicago, published a periodical, and set up a commune called Koreshan Unity with 4000 members. However, a new Jerusalem which he set up in Florida fell short of expectations as the anticipated eight million members turned out to be a mere 200.

Teed, or rather Koresh, the man who attempted to transform the Hollow Earth doctrine into a new religion, died in 1908 after promising his followers that he would materialize after death and escort the entire community of believers to Heaven, finding his postmortem return unavoidably delayed, Teed's followers buried his body beneath a huge monument on Estero Island. During a great storm, a huge tidal wave swept across the island shattering the tomb to pieces and bearing away the body of the 'Messiah' into the ocean's depths. A fitting end one might think to Symmes's Hollow Earth theory.

Icy Stars

However, this was not to be. Symmes's ghost refused to be exorcised and was to haunt mankind in the most unexpected way. A German ex-World War One airman named Peter Bender was attracted to the Koreshanity cult and popularized the theory, thereby influencing many early members of the National Socialist Party of Germany, where it was called the Hohlweltlehre or Hollow Earth doctrine. In Jacques Bergier's and Louis Pauwel's book *The Dawn of Magic* we read: "The Earth is hollow. We are living inside it. The stars are blocks of ice. Several moons have already fallen on the Earth. . . . Such are the scientific theories and religious conceptions on which Nazism was originally based and which Hitler and his supporters believed."

Symmes lives on in literature, politics and science fiction and it seems somewhat sad therefore that the psychologists should have written off the entire Hollow Earth theory as no more than a sublimated desire to return to the womb.

Symmes "No. 1. Circular"

LIGHT GIVES LIGHT, TO LIGHT DISCOVER -- "AD INFINITUM.

ST. LOUIS, (Missouri Territory,)
NORTH AMERICA, April 10, A. D. 1818.
TO ALL THE WORLD!
I declare the earth is hollow and habitable within; containing a number of solid concentrick spheres, one within the other, and that it is open at the poles 12 or 16 degrees; I pledge my life in support of this truth, and am ready to explore the hollow, if the world will support and aid me in the undertaking.

Jno. Cleves Symmes Of Ohio, Late Captain of Infantry.

N. B. -- I have ready for the press, a Treatise on the Principles of Matter, wherein I show proofs of the above positions, account for various phenomena, and disclose Doctor Darwin's Golden Secret.

My terms are the patronage of this and the new worlds.

I dedicate to my Wife and her ten Children.

I select Doctor S. L. Mitchill, Sir H. Davy, and Baron Alex. de Humboldt, as my protectors.

I ask one hundred brave companions, well equipped, to start from Siberia in the fall season, with Reindeer and slays, on the ice of the frozen sea: I engage we find a warm and and rich land, stocked with thrifty vegetables and animals if not men, on reaching one degree northward of latitude 82; we will return in the succeeding spring.   J. C. S.

DR. MTICHILL TO CAPTAIN SYMMES.

The following letter has appeared in the public journals, and we believe it may be relied upon as an authentic production from the pen of "one of the men, who honor America most by his information and talents;" and who has "a great share in the new glory which awaits our country." The letter from the explorer will be found (ante p. 445), and we thought we had saved the learned professor the trouble of writing a reply, by our voluntary communication on this important scheme, (vide ante [Feb. 1818] p. 123.) But some men will manage the own affairs in their own way. The doctor is a worthy old gentleman, and whether he encourages the wild adventures of Symmes, or flatter the "dear girls" of New York, we believe he means no harm to any body. His first object is to gratify a most inordinate vanity, but in seeking the means of administering to this passion, it must be admitted by all that Dr. Mitchill has done the state some service.

New York, 16th June 1818.

Sir – The important enterprise sketched in your letter lately received by me from St. Louis, brings to my recollection several facts and occurrences relative to the polar regions of our planet.

You doubtless know the zeal and perseverance with which our countryman John Churchman, urged to Congress and to other bodies, the importance of a voyage toward the North Pole. His object was to find the magnetic pole of the earth, which he affirmed to be several degrees from the axis on which it seems to revolve. But he did not live long enough to prove his doctrine, nor to ascertain the revolutions of his magnetic poles around the two extremities of the globe's axis. I remember him very well. His book is extant.

The departure of the ice in vast masses from the arctic regions began to excite general attention in 1805 during that year.

[p. 447]

I investigated the subject, and wrote a memoir upon the Greenland ice, which overspread the northern Atlantic Ocean, and cooled the water and atmosphere enough to be felt in our climate as far south as 40 deg. north. I consider the Gulf Stream as acting by its current to carry the ice away to the eastward, and by its warmth to melt it. Thereby this marine river saves the bays and harbors of our coast from obstruction and blockade by these congealed masses. This easy, with the testimony of many ship masters, is registered in the tenth volume of the Medical Repository.

THE BEST OF THE HOLLOW HASSLE

A few evenings ago, Captain White, now of New York, told me he had, in the year 1774, penetrated on a whaling expedition as far as 82 degrees 30 minutes north. He was encompassed by floating fields of ice. The water of the ocean frequently curdled or thickened to icy crystals between them. The ship's rudder was unhung and taken on board, as being of no use; and the needle of the compass became torpid, or sluggish, to such a degree, that there was a necessity to shake the card, for rousing and waking it up, as it were.

I wish success to the enterprises of the English for visiting once more the high latitudes. It would be gratifying to me that the inhabitants of our continent, which reaches very far to the north, should be foremost in exploring its extent and boundary. Men of ardor in the cause, and of hardy resolution, and of prudent foresight, are the proper persons for engaging in such adventures.

There have been various speculations, on the constitution of the internal nucleus, or core of the earth; some considering it as occupied by solid rick, others by water, and others again by fire. Ulysses is represented by Homer as penetrating to the nether abodes by the way of Cimmeria – and Aeneas is said by Virgil to have descended to the lower regions at Avernus. Dante has given a map, or profile, of the spaces between the crust of the globe and its centre of gravity, as an embellishment to his poem Inferno.

But all these are visions of the imagination, of fictions of poetry; we stand in need of better information; one actual explorer would be better than a thousand inventors of stories.

[p. 448]

How rare and extraordinary would it be to converse with you, on your re-appearance from the internal worlds! I told Captain Lewis and captain Riley, on the return of the former from the northwest coast of America, and the altter from the frightful deserts of Africa, that I beheld them as, in some sort, visitors from another sphere; so would you really be after the performance of the project contained in your letter. Adieu, and be happy!

SAMUEL. L. MITCHELL.

John Cleves Symmes, Esq.

[p. 471]

21

THE BEST OF THE HOLLOW HASSLE

Captain Symmes again. – Captain Symmes' theory of the earth is not quite so novel as is generally thought; the idea of the globe being hollow at the poles was suggested many years since. In a work published in Paris by an anonymous writer, called "New Conjectures on the Globe of the Earth," the author asserts, "that in examining the internal parts of the globe, it is not possible to doubt, but it is a composition of several beds of slime arranged upon each other by the waters of rivers, and consisting of the substances which they contain, and which these rivers carry off from the rising grounds, in order to deposit them on their banks, or in the bottom of the sea, to which they run; that the globe of the earth was originally formed of a flat crust, composed of these depositions; that this crust being very thin (only two thousand three hundred and eighty fathoms thick) includes a very subtle air, is supported by the weight of a double atmosphere which surrounds it; that this equilibrium having ceased at the time of the deluge, the crust was broken and scattered; that its wrecks floated in the sea as clouds do in the air, and were heaped on each other, and in certain parts so accumulated as to form certain prominences; that our mountains proceeded from this; that by this subtraction from the crust of the earth, of the prices by which the mountains were then formed, there remained vacuities in this crust two or three hundred leagues in diameter; that it is by means of these apertures that the seas of both surfaces of the crust, at present communicate with each other, that these seas enter by the poles into the cavity of the globe, and turning round this cavity in a spiral line, they come out between the tropics, and causes the flux and reflux of the sea, which are some sensible in one part than another, according to the position and largeness of the passages through which these seas enter or come out."

From the *National Intelligencer.*

Miscellany.
Cincinnatti, Jan. 18th, 1818 [sic – 1819?]
Messrs. Gales & Seaton:

Pope advises authors to keep their works many years – I correct mine as often as I peruse them – hence cannot instantly profit by your acceptance of the offer I made of sending my new

22

memoir for publication. My progress in philosophy is voluntary or spontaneous, and not the consequence of immediate volition; so hurry suits not with my studies: my intentions, however, seldom subside until accomplished; hence you may depend on shortly obtaining the memoir alluded to in my last -- and, in the mean time, I add below, some of the particulars to be explained in future numbers.

My family requires most of my time and efforts. I shall not, however, neglect to develop my new principles, even though it should cost a portion of the patrimony designed for my children. If the world, or some national governments, do not furnish the means to explore, as I have asked, I can proceed but slowly with my investigations, for my pecuniary concerns have been so much neglected lately, that I shall have to lay aside, for a time, several new memoirs in a progressive state, including one on the source and production of animal free heat. Wishing my writings to be as free as air, I am unwilling to put them to sale: indeed I should prefer that my pupils, like those of Doctor Black, should themselves develop my discoveries. Besides the time expended on my new positions, I have paid out considerable sums for the printing and postage of five hundred circulars, of which I distributed one to each notable foreign government, reigning prince, legislature, city, college, and philosophical societies, throughout the union, and to individual members of our National Legislature, as far as the five hundred copies would go.

I have much to say, but will conclude my address with only a quotation from Nicholson's Encyclopedia under the head "Earth." "The attentive and skilful observer of the works of nature, whether employed in examining the most wretched or the most sublime, will find that judgment, and infinite wisdom and ingenuity, has equally prevailed throughout. Can it be supposed for a moment, that the internal part of the earth we inhabit has received less attention from the Creator, than those objects which are under our immediate and unimpeded inspection."

Respectfully,
JNO. CLEVES SYMMES.

Light develops light from age to age.

The data I have as yet obtained, indicate, 1st. That, the axis of the earth is not in the centre of the polar opening, but several degrees towards Spitzbergen or Siberia.

2. That the magnetic needle regards the centre of the polar opening, rather than the axis.

3. That the needle should so turn, on entering the polar opening, as to have the same end S. within that is S. without.

4. That (contrary to my former idea) this sphere northwards towards the polar opening, is rather a protruded sphereoid than a depressed or oblate one.

5. That much of the water, developed to air or vapor within our tropics, is condensed to abundant rains by the increasing gravity towards the internal equator, thereby setting free latent heat and light.

6. That the haze or smoky appearance of the Indian summer comes from within the sphere, although S. winds often thicken it by heaping it upon itself.

7. That the northwestwardly winds generally come out from some of the poles of the inner spheres, and that the northwestwardly winds come from the concave surface of this sphere, as do the southeast monsoons.

8. That the polar opening is the source whence the matter of our snow storms is derived -- although the snow be crystallized on, or after its passage over the icy hoop, or circle.

9. That when the sun is 23 degrees south of the equator, the line of the greatest cold north is 23 degrees S. of the polar opening, or ninetieth degree – and when he is 23 degrees north of the equator, this line of cold is removed 23 short degrees, beyond the ninetieth degree.

10. That the dark complexion of the nations high north is derived from the hot climate beyond.

11. That, when 90 degrees real, refraction will so deceive, as to indicate eight or ten degrees less than 90, owing to the atmosphere extending over the polar opening – so as to cause the zenith of the atmosphere, at 90 degrees real, to be depressed at a considerable angle towards the south. For example, when a person traveling north, has brought his horizon at right angles with the

plane of the polar opening, he will have the zenith of the atmosphere there, nearly or quite in his horizon – he must, therefore, lose about 20 polar degrees in estimating his progress, if he judge by celestial observation only.

12. That a part of the sphere near the verges, is water quite through, so that it transmits light within.

13. That the many large floating trees found as high as 80 degrees north, which is 20 degrees beyond where we find such grow, float out through the polar opening, and wedge in the broken fresh water ice that surrounds it -- which may be called the icy hoop or circle.

14. That the spheres high north are thin, as gravity there is found but little greater than at the equator, although the centrifugal force amounts to almost nothing.

15. That clouds, haze, or mist, will generally prevent a view of the opposing spheres, or polar verges.

16. That a murky atmosphere of mist or haze hangs over or about the north polar opening, sufficiently dense to project a spherical shadow on the moon when she is eclipsed.

17. That beyond north latitude 75 degrees, when the sun is seen near the northern horizon, he must appear much higher than he really is, owing to the refraction of the atmosphere -- and it is this extent of atmosphere which makes him look (while so situated) dull like the moon.

18. That mackerel, cod, whale, and the musk ox, inhabit for towards the internal equator -- as the first and last breed when absent, and as the fat or flesh of each readily develops (when without the sphere where the centrifugal motion lessens the force of gravity) to a more fluid or volatile rancidity than is common to local fish or animals.

19. That the polar verges yield to the gravity of the moon, so as to effect our weather at her changes, from the air being either sucked in or forcibly protruded by such action.

20. That tornadoes proceed from a convulsed disruption of the first aerial sphere above us, through which is forced down a gush of confined elastic fluid.

21. That the blaze or fire of our volcanoes is the heat set free from a latent state, when the elastic fluid of the mid-plane of the sphere is forced up to where the greater gravity of the surface condenses its molecules to their original stony base at Vesuvius, and watery base at Hecla.

22. That along the mid-plane of the solids of each planetary sphere, there is a place widely filled with elastic fluid, and distended with fluid molecules to a limit and rarity, greater in proportion to the greatness of the gravity of their external or exposed surfaces – and thus they serve to buoy the earth, comets, fireballsm and all planetary bodies, as balloons are buoyed. This principle (to borrow the language of another) has, I think, "the advantage of simplicity, and simplicity the offspring of unerring wisdom, and Almighty power, is in geenral, the companion of truth."

It has been asked, how should I be able to make farther discoveries than others? I answer the question by another, why not – when I have all the prior ones formed, whereon to found my new ones?

I make a general request or invitation to editors, that they insert this miscellany in their periodical journals, and also my other writings as they may occur.   J. C. S.   (text taken from a reprint of the National Intelligencer item, printed in the Gettysburg, *Pennsylvania Republican Compiler* of Feb. 2, 1819.

From the *National Intelligencer.*

ARCTIC   MEMOIR.
Cincinnatti, Feb. 28, 1819.
I hoped ere this to have been supported in my new theory of the earth by many pupils, but find that most of those who have written are inclined to oppose me. I would prefer having an advocate to state my views, because in proportion to their extent, I may subject myself to the imputation of extravagance or ostentation, especially as while I write, I naturally feel elated with my discovery. I am, perhaps, better fitted for thinking than writing –

reared at the plough, I seldom used a pen (except in a common place book) until I changed my ploughshare for a sword, at the age of 22, not wherewith to carve a fortune, (having already an ample farm by the liberality of my revered uncle after whom I was named,) but to merit and obtain distinction, and accumulate knowledge, which I had seldom tasted but in borrowed books. – With respect to the latter the world is now to judge of my success; and in relation to the former, I at least may say I satisfied myself and fellow soldiers, if not my country – not only at Bridgewater on our left, and the sortie of Fort Erie in the van, but throughout my 13 years' service, ending with the war.

I presume few have inquired more devotedly than myself into the reason and origin of all that occurred to view, I remember, when at the age of 11, (in Jersey) while reading a large edition of Cook's Voyages, my father (though himself a lover of learning) reproved me for spending so much time from work, and said I was a book worm: about the same age I used to harangue my playmates in the street, and describe how the earth turned; but then, as now, however correct my position, I got few or no advocates. I must not, however, say I get no advocates, for I have several. I particularly boast of two ladies, of bright and well informed minds, on the banks of the Missouri, who are able and earnest advocates, and devoted pupils; to them is sue the credit of being the first to adopt what the world is so tardy in admitting. But, Col. Dixon, who has traded on Lake Winepec, with the Indians, ism I presume, the most impostant pupil I have obtained, for he has long been actually engaged in the N. West Company and fur trade. He declared, in our first interviews, that I was certainly correct, and stated to me many important, otherwise inexplicable, circumstances, occurring high in the north, that were completely solved by my principle: he is regarded by such as have long known him at St. Louis, as a gentleman of a very strong and well informed mind.

In addition to the passive concurrence of several men of thinking minds, among them a venerable member of the American Philosophical Society, in this neighborhood, I have been honored with the offers of several more enterprising spirits to accompany me on the expedition I propose; but as the conditions with regard to my outfit by the world, are not yet complied with, I have not positively accepted of their services. I still hold my life pledged, however, for the general truth of my position, and devoted to the exploration. I calculate on the good offices of G. BR\ritain and France, for they nurse and patronize the sciences with ardor. My wife boasts her descent from the latter, and I trace mine from the former. FRom the Emperor of Russia, so well known as a patron of scientific enterprize, I flatter myself with much support.

I challenge any opposers of my doctrine, to shew as sound reasons why my theory is not correct, as I can shew why it is.

I refer to those who seek for truth to Ree's Cyclopdaeia, and any other books wherein the quadrupeds, fish, and phenomena of high latitudes are treated of; likewise those books that treat of Venus, Mars, and Saturn -- where they will find many tests, that if duly considered, must go to prove my position.

In the Cyclopdaeia, under the heads Fishery, Arctic, Herring, Seal, and all the other migrating fishes, it is shewn that most or all of them retire annually beyond the icy circle during the winter, and return, increased in fat an numbers, in the spring; and, under the head Reindeer, it is stated that this animal passes annually near Hudson's Bay, in columns of 8 or 10,000 from N. to S. in the months of March and April, and return N. in October, as stated under the ehad Hudson's Bay. I propose to follow the route taken by the reindeer, northward in Siberia, where they depart every autumn, from the river Lena, (as Professor Adams, of St. Petersburgh, states,) because it is probable these deer choose the best season and nearest route, to fertile and habitable lands. I propose returning either in the course of thorty or forty days, or when the columns of deer return in the spring. It is presumable that men can live where deer thrive. I dp not think there are no dangers attendant on such a trip, but believe the object will justify risking all probable ones.

In plate 17, vol. 33, part 2, of the Cyclopdaeia, the figure of Mars, with his equator towards us, exhibits his poles, surrounded with single light circles, whose farther sides extend beyond the periphery of his disk. I hence conclude that his poles are open, and that the light reflected by the farther sides of the verges of the openings, is refracted so as to appear extended beyond his disk, by means of its coming to us through the nearest verges. It is a well known fact, that refraction is greatest towards the poles, owing probably to the dense atmosphere there. The apparent continuation of the margin of his true disk through these rings, (if not an imaginary line dotted there,) must be the farther verge of the second second sphere within, rising by refraction, apparently as far out as the periphery of his disk.

I contend that the space within the circumference of the arctic icy circle, if not hollow, or greatly concave could scarce afford space and surface to maintain alive, and in health, all the fish known to come from thence annually, in the spring, even if (without resorting to feeding on each other) their food was inexhaustible, anmd the whole circle water. But floating trees being often found far north of where we say any grow, is an impressive circumstance to shew it cannot be all water; and the fact that these trees are generally such as abound in the tropics, (together with several unknown species) shews that there is a hot climate beyond; and the migration of the reindeer, too, shews that moss or other vegetables abound there, and consequently land. Pinkerton shews that the Dutch, who at different times got detained by the ice, could find but few fish to eat in the season of winter.

I also refer to Doctor Darwin's note on winds, in his Botanic Garden (which I never read until after I asopted my theory,) ehere that great although often extravagant philosopher declared his belief that there was a great secret, yet to be explained, at the poles, and anticipated that the light of the present age would disclose it. The stone spheroid he found hollow and disposed in concentric strata, and the concentric iron nodules he describes, deserve to be considered. He states that the seeds of several tropical plants are often found in the seas high north, in a state so recent as to vegetate.

I recommend the perusal of Mavor's and Pinkerton's voyages: Pennant and Goldsmith on animated nature; and [Heald's] and Mackenzie's travels -- wherein many tests of my position exist. Pinkerton shews that beyond latitude 75 degrees, the north winds are often warm in winter; that in mid-winter there falls for several weeks, almost continued rain. and that vegetables and game are more abundant at 80 degrees, than at 76.

When my chain of reasoning (drawn from the nature of matter) first led me to the conclusion of hollow spheres, and open poles, I merely intended broaching it as a question; but, when I found the planets of the heavens, and the phenomena and natural history of the polar regions, afforded proofs incontestible, I then seclared the fact without reserve; and have been considered by many as a madman for my pains. Were I to feel in any degree disconcerted by the playful, however ill-timed witticisms of others, I should comfort myself in the reflection, that, so soon as I shall succeed in the establishment of my theory, the more it has been decried, the more I shall feel honored in the event: innovations in science or art, most commonly exite opposition.

If additional reasons are required, I have adopted ab ample fund yet in store for the world.

JNO. CLEAVES SYMMES.

(Appended to this Memoir is a Note, which Mr. Symmes writes us need not be published, unless it be convenient to have three diagrams cut to accompany it. It not being convenient, we are obliged to dispense with the publication of the 'Note,' which is illustrative only of Mr S's theory, with reference to a diagram representing the ideal "section of a nest of spheres cut through the poles, as an outline fo the formation of the earth," and to the telescopic appearance of the planet Mars, shewn in the Cyclopdaeia and in Ferguson's Astronomy, as confirmatory of the theory of polar cavity.

Editors Intelligencer.

(text taken from the April 7, 1819 issue of the Gettysburg Republican Compiler)

CHAPTER THREE

Our Paradise Inside The Hollow Of The Earth
By Tal LeVesque
VOL. 2 NO. 1 – OCT. 1980

"He stretcheth out the north over the empty place." - Job 26:7 "I will sit also upon the mount of the congregation, in the sides of the north." - Isaiah 14:13

Over 100 Bible verses teach that the world is Hollow. These verses speak of a place UNDER the outer shell of the Earth. Every verse that speaks of UNDER the Earth refers to Paradise in the Hollow Earth. A CENTRAL SUN lights up the whole Interior of the Earth.

The whole interior of the Earth is called "Eden," the garden of God. The word "Eden" means "Paradise." It is a grand enclosure or garden, out of which our race came from.

Paradise Lost & Paradise Restored, we are on our way back to Paradise inside the earth. We shall fulfill the covenant; when we go back into the garden; into the abode of LIGHT, wherein there is NO DARKNESS (i.e. No Night). You see, the present surface world is not our home. We will inherit our true home land...inside the Earth. In this lovely Paradise there is a Sun, a glorious Sun that shines with PERPETUAL LIGHT. It is a stationary Sun that never moves, never rises, and never sets. In this glory land it is one continuous day, for there is no night there.

From history and tradition we learn that there are several SUBTERRANEAN PASSAGES or Great Wide Tunnels which lead from the exterior to the Interior of the Earth.

Pindar, the Greek poet who lived from 522 to 443 B.C. said: "Happy the man who descends beneath the Hollow Earth, having beheld these mysteries, he knows the end, he knows the Divine Origin of Life."

THE BEST OF THE HOLLOW HASSLE

Plato writes, concerning an Interior Sun, Apollo of the Inner Earth: "He is the God who sits in the Center, on the Navel of Earth and he is the interpreter of Religion to All Mankind."

In ancient Pagan times the superior race who inhabits the earth's interior frequently contacted surface dwellers, and temples were built for their occupancy during their visits. But with the establishment of the Church of Rome, these temples were destroyed and the existence of the "gods of the Underworld" was denied. Christian priests taught their followers that they were devils, evil beings who should be avoided, and that they should worship only, as the Church-State, who "wanted to control them, taught.
Thus the existence of a Subterranean World and its inhabitants became a lost memory. And just as religionists taught that it was an inferno of everlasting fire, so scientists preserved the error in their theory that the earth has a fiery core, basing this on the flimsy evidence that the further down one goes, the hotter it becomes. However, this evidence has now been proved false and scientists are looking more closely at the Hollow Earth theory, as they find that most things in nature are Hollow.

Every prophet, however great, must be initiated, Christ and his mother were members of the Essene (is-sonir - sons of ice} community. As an Essene priest, he could not have instituted the neo-Christian sacraments attributed to him. On the contrary, he would, have restored the ancient Essene-Druidic sacraments. Christ is a Cosmic Power, the Chosen One of the Devas, the Solar Word, and not the dogma of organized religion, of Churches today. DOGMA reversed is AM GOD!

THE MYSTERIOUS WORLD OF CAVES

CAVERN NEAR PITTSBURGH - "AMAZING STORIES" Somewhere between 12/46 & 6/48
Letter sent in by George A. LeHew, 1918 W. Newport Ave., Chicago, Ill.

I too, know of one of these caverns into the world below. It is about
fifty miles south of Pittsburgh, Pa., in the first range of the Allegheny
Mountains. My experiment with the caves have been only partial explor-
ations, consisting of traveling about a mile and a quarter down into the
cave itself, and returning. The cave is ventilated from below, and
stays at a constant 50 degrees no matter what the outside temperature
may be. It is a series of rooms or galleries with narrow passages from
one to another. In about the sixth room down there is a large tree
trunk which could not have come from the surface above as the stratosphere is almost completely free from local fault; and it could never
have come down through the openings in the cave itself as they were
small at the top, and kept getting progressively larger as they got
deeper.
I traveled down as long as I could find comparatively easy travel -
about a 45 degree descent all the way - and finally came to what I
thought must be the end of the cave, for I could see no more openings
into rooms, but on closer examination found instead a bore, about six
feet across, straight down into solid rock. I turned my flash downward
and could see that it must have gone straight down for at least a
hundred feet, the sides were perfectly smooth, and the shaft, or bore,
in a perfect round - no apparent irregularities anywhere. I had no way
of descending any further, so I returned my steps back up through the
different rooms to the top of the mountain where the cave opens to this
world. I made discreet inquires of several old timers in that region,
and found that in 1915, or about that year, six survivors took gear and
equipment, and spent a month in exploration of the cave, going 18 miles
from the entrance, and down almost five miles below sea level. I have
never gone back, but hope to some day in the future, with escort, equipment and supplies. I'd certainly like to see the maching that made that
bore! Oh, yes, one more interesting item - the surveyors in their exploration of the cave, distinctly heard the rumble of machinery - but
their calculations proved they were nowhere near a large city, (surface)
and they were too deep for surface noises otherwise.

CAVES OF BONAIRE - Source unknown

Cave explorers wearing scuba diving equipment have discovered a mysterous
staircase, obviously ancient, leading to a network of caves on the
Carribean Island of Bonaire. There are 18 steps in all and it is believed
they could not have been carved by the primitive island inhabitants who
lived on Bonaire when the Spanish arrived there in the 15th Century. The
cave explorers, let by Don Stewart, manager of the Flamingo Beach Club on
the island, found no artifacts in the one cave they explored but reported
that the walls and ceilings were covered with red and black inscriptions
which "resembled" Mayan hieroglyphics. There are at least 25 other large
caves on Bonaire which have not been investigated - by modern man.

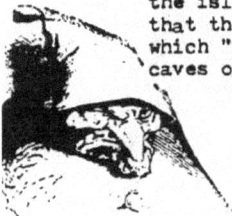

CHAPTER FOUR
How Things Really Look
By Richard S. Shaver
VOL. 4, NO. 4 - Aug. 1983

It is frightening to sit and watch the regular TV programs while at the same time reading an issue of *Saucers, Space, and Science* or *Probe* or *Phenomena Magazine*...to get the perspective of the whole world outlined against the picture of swooping saucers quite ignored by that world of ours.

If you remember the "War of the Worlds" by H.G. Wells...as he draws the picture of descending invasion ships...the parallel to our present series of flaps if really startling.

Not believing in flying saucers is very like not believing in lice, or bedbugs, or any of those unpleasant hiding things we like not to think about. You have to take measures. The Great Lakes trout probably didn't believe in lamprey eels until they got them on their backs, sucking the life out of them.

Most UFO buffs, from editors of zines down to the lowliest casual reader of UFO zines, have one hope on their horizon. That is that their efforts will lead to some sensible attention by the authorities to the broad problem the UFC present...the problem of "are they invaders?" "Why do we have alien ships in our skies?" "What do they want here on Earth?" And, "are they really from space?"

To people like me, who have been with the modern UFC scene since long before Arnold's sighting of saucers hit the newspapers, this picture of worldwide disinterest is the most frightening aspect of the saucer scene.

To me, this is the same phenomena as the soporific a mosquito injects before it bites. It is tactical, designed and carried out by an immense organization who is in fact taking over our planet. The device they use to apply the soporific is the one I have written about as the TELAUG in many of my stories. TELAUG is short for telepathic augmentive device.

People and government officials are myopic about UFO because their minds are affected by the telaug. When a saucer hovers over a car for a time, then takes off in a burst of speed...there is only one reason for the hover. No one but myself has ever given the reason any credence.

The reason a UFO hovers over someone is very plain to me, but to no one else, apparently. It hovers to get its bearings from the mind of the local inhabitant...who carries his directions and his knowledge of the locale in his head. The UFO doesn't need to go to the service station and buy a map...he picks it all up out of the nearest MIND.

The broad perspective of the past space traffic of the planet Earth is needed to appreciate the source of the UFO, If you read the rock books put out by the past races of mankind which litter our hillsides, (ignored even more than the UFO), you would know there are a whole string of planets out there in space from which the UFO come, and to which they go.

They come to Earth for things surface men used to know about but which we have lost knowledge of in late centuries. They come for rock books, for one thing. And they also come to enter cave cities we don't even know exist...to buy antiques. They come for parts for their ancient saucer ships...parts once manufactured here on Earth in our underground factories.

We must look very pitiful and helpless when they probe our minds for directions and find out we don't even know the telaug exists, or that Earth once manufactured saucer ships or anything about the ancient space routes they followed to reach Earth.

Pitiful, helpless, victims of ignorance...must be their thoughts as they speed off. Unfortunately, I suspect their thoughts are more like the thoughts in the mind of a large lamprey eel as he selects his next victim from the nearly extinct trout of the Great Lakes... not thought at all but a victim sort of greed.

For the ships they drive were manufactured by our own race, long ago before the deluge plunged Earth into the darkness of worldwide ignorance. Their science is not their science, but our own, long ago. They don't have any, to speak of.

How to get across what I know about the UFO and the broader aspects of the UFO invasion of our skies, our oceans and our underworld (of which we don't even know) to the average man, has for me always presented an insoluble problem.

I know...but I can't get what I know accepted as fact, anymore than any UFO sighter can get his story accepted as fact. There is always the element of doubt, even in photos of UFO's, and many a UFO photo is suspected as a fake.

Though I know the telaug operates in all our cities, and is behind such horrors as the Kennedy murders, the King murder, The Dag Hammerskjold murder, etc., I have found it impossible to get the picture accepted generally as the UFC themselves are accepted.

Today, now...I have pictorial proof quite as good as the UFO photos of saucers in our skies...but I don't expect any greater official acceptance of their factual nature than the UFO photos get.

The reason for official non-acceptance of UFO data is the telaug...holding the official mind into its rut of conservative insistence upon "more data," "more proof," and the official preference for explanations in place of action.

The truth is that asking officialdom for action about UFO data is very like a Jew asking the Hitler regime for mercy on the grounds that he has a family to care for. The storm troopers reaction is to go out and pick up the rest of his family. It is also like one Great Lakes trout asking another Great Lakes trout for some action about the lamprey eels. The fact is they both have a lamprey on their backs.

That lamprey is the telaug, operating them like a puppet is operated with strings. They CAN'T do anything about anything unless the puppeteer wants them to do it.

We are lucky we can talk this much about it. Not that such talk really does anything about the problem. Except to give the next UFO better bearings on the problem when he picks up his bearings from your mind.

Earth will never understand or accept the UFO invasion until it learns all about telaugs and adapts them into the life of the surface world as the radio has been adapted into our life.

I am really afraid there isn't much hope for us. We can't get anything much done for our own welfare as long as our leaders like Kammerskjold and JFK are murdered regularly. A leader isn't much good to us dead.

What really worries me in this leader picture is that leaders like Stalin and Hitler live on and on, leading us all to death...while the good ones like JFK are murdered as soon as they get the reins in their hands.

This is the broad perspective of the UFO invasion...a steady insertion of murderous tactics and complete repression, via the telaug, into the picture of surface life on Earth.

The world wars present the same sort of picture when you think of them as really PUPPET wars, very like the Pied Piper of Hamelin, leading the kids into oblivion.

Our world wars and our Biafra situations aren't really what they seem, you know. They are really the broad destruction of all forward growth in our society...to be replaced by utter mind slavery!

Mind slavery is the picture the UFO in our skies present! That this mind slavery is not new on Earth doesn't make the picture anymore inviting to one who knows what is going on. The people on Earth, to me, are like a blind man in the coils of a python...he does not even know what is killing him!

We are like the lake trout of the Great Lakes...and we can't do a thing about the creatures that are killing us...just as they killed the Kennedy brothers and Dag Kammerskjold and a million others not so renowned...everyone of us who puts out his hand toward the door of freedom of mind.

Freedom of mind, freedom of choice, individual mental existence is what is at stake! And you won't win that freedom without the telaug, and you won't win to space without the rock book plans for the building of your own space ships....

THE PROJECTIONISTS
By Richard Shaver
VOL. 5, NO. 1 - Oct. 1983

I know that a lot of UFO sightings are really only projections, there is really no saucer present, however, that is not always true...but when a genuine UFO is present... it was in ancient times customary battle tactics to use a number of solido-graphic projections as evasive tactics. In fact, the way ray weapons are built they don't miss...and the only practical evasion was this duplicate projection in numbers.

So...FLEETS of UFO's when seen are as likely as not ONE real UFO hiding among a series of projections.

The whole subject of ray phenomena as well as UFO phenomena...is in fact PROJECTION work...just as ghosts boil down to projection work...and most psychic phenomenon turn out to be projection work when investigated capably...often with modern devices.

BUT the mechanisms used to make these projections in 3-diraensions are themselves related to the UFO phenomena in that they are pre-deluge in origin... the product of past civilizations long vanished.

Corroboration for this can be obtained by studying Fort's chapters on this type of spectral phenomena...and the same is true of many accounts of psychic phenomena.

Another fact that bears out the projection (fact) is the impossibly tight turns. Often at right angles at high speed...the g's would crush any normal pilot. Only a projection can make this sort of maneuver...but the average UFO observer accepts it as a genuine solid object...which it is NOT... it is a projection.

So it is with some faces in photos...there are always some in nature in all photos...but they are of two kinds. ONE...natural apparitions called "Accidentals"...and unnatural appearances that are in fact projected with the antique mechanisms from the buried cities underground.

If you still haven't accepted the facts...a study of this angle of UFO phenomena should help to convince you of the spectral or projection source of many UFO sightings...same is true of psychic stuff.

Cloud pictures...are sometimes accidental and sometimes are projections ON the clouds.. .of such a nature they seem to shape the cloud and paint with the cloud material itself...and I have often talked with underground people by watching clouds and what they have to say they illustrate with cloud paintings as they go...this cloud painting was in fact a favorite pastime of the ancient peoples and such devices stand about in the buried cities waiting for someone to use them.

The whole panorama of "phenomena" is so inter-related with the far past and the surviving apparatus in the hermetically sealed cities (most buried cities were hermetically sealed by floods of sea bottom mud) is the reason the metal and perishables survived the ages of time that passed...they are hermetically sealed by the mud flows.

Just like a tin can...many things survived until today because protected from all corrosion by the mud seal.

Thus the food supply of the billions of people who inhabited the buried cities...was left behind in storehouses and hermetically protected from decay quite by "accident"...due to the rivers of mud dredged up by the mile-high tidal waves and deposited everywhere...which few people today understand... and is the real food supply down there.

Stored food meant for multitudes...was never used after they evacuated, and was protected hermetically by chance...since broken into this ancient stored food is the base of underground livelihood.

So with saucers...and so with projections from ancient devices...it is mere chance that protected it all from time's corrosion.

Just as jars of honey in the pharaohs tombs were still usable...and just as grains of wheat brought out of tombs still sprout...so in the long buried cities many things survived that one would not expect to find.

So don't think UFO and UFO sightings are separate from the Shaver cavern "theories" (which are not theories but facts I know) because you can't separate them...they interlock.

SOUTHWEST EXPEDITIONARY UNIT
A PRIVATE UNIT OF
SPELEOLOGISTS, ADVENTURERS & EXPLORERS
Charles A. Marcoux - Dir.
P.O. BOX 11494   PHOENIX, AZ. 85061   USA

MARY DAVIS
MARTIN

Charles & Lorene Marcoux

DOLORES CROUT

ACTIVE EXPLORERS

DENNY DAVIS

LA BRON BYNUM

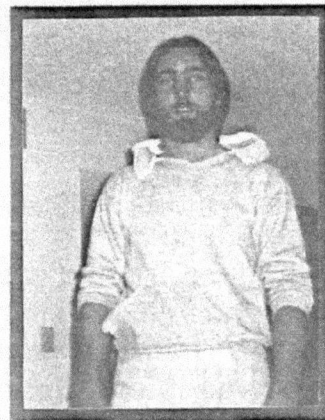

WILLIAM CROUT

PAGE 1

DOING THE IMPOSSIBLE   IS OUR BUSINESS

CHAPTER FIVE
I Live With The Teros
BY George D. Wight & Charles A. Marcoux
VOL. 3 NO. 4 - Aug. 82 – VOL. 4 NO. 1. - Oct. 82

This is straight talk, not sweet talk, nor is it from the horse's mouth, or is it a long shot on the horse's nose. In the past issues of the Hollow Hassle, I have answered many questions about the hidden secret caverns of the ancients. In many ways, I have tried to encourage your endeavors on these subjects, and at the same time, have warned you to be cautious during your various adventure trips. Therefore, the Hidden World is not all "sweet talk," but "straight facts," and your search for an entrance into the "Inner Earth" by way of the secret hidden caverns is a "long shot." I am also aware that some have intentions that are not beneficial for the welfare of the hidden world, the secret people or the sincere, surface people. For various reasons of my own, I have been unwilling to reveal certain information that will help you in your search, and the sincere people must pay the price, in order to keep this information out of the hands of the destructive, criminal-minded people. Believe me, there are people who "prey" on us and the various "zines" for the information received from sincere ones, and they (including government agencies) use it against us.

FOREWORD: In this issue I will present information to encourage you to go on your own search and, maybe, you'll have the opportunity to join my expedition into the hidden caverns, with the possibility of personally meeting the secret people who are known by many as "Teros." In this article, and in future issues of the *Hollow Hassle*, I will reveal various material and information that will give you a completely different concept of the cavern world and the way of life of its people who have lived there for centuries, since the fall of Lemuria and Azaltan (Atlantis). I'll also explain some of the taboos and remove many doubts from your mind.

At the present time, I will not divulge the names of people involved in Mr. Wight's experiences with the Teros, nor will I reveal any information as to where this certain cavern is located. This is done in order to protect all concerned, including the Teros. Since this is being written from notes, like a personal diary, I have had to sort out the material, put it in sequence and edit it for a series of articles, so that the readers will be able to understand the material and put it to use for their own investigation. Now, read on, as I guide you into this greatest adventure of your life and into the "LAST FRONTIER."

THE LAST FRONTIER

I am sure that many of you have seen the old movie, "Lost Horizon," starring Ronald Coleman. It is a story about immortality and a hidden valley somewhere in Tibet. The movie's story is based on legends, myths and superstitions about a "Hidden Kingdom of Shambhala," which is a sacred city in the bowels of the Earth, and it is ruled by the "King of Kings," who is ruler of this Earth – so goes the legend. Then, do you remember an old movie titled "SHE" with a queen that ruled a city built in the caverns, deep below the earth?

Also, do you recall the movie, "Journey to the Center of the Earth," which gives you some ideas of the cavern world? Even though all were portrayed in fiction, they were based on the many legends, myths and superstitions of the "Underworld." The last two were places of demons, ghouls, monsters and whatever else was conceived of as fearful by the superstitious natives throughout all parts of the world.

Still, others spoke of wonderful Gods, giants and supermen who warred against the satanic devils in the various concepts of Hell in bottomless pits, where Deros or fallen angels tormented all the saints and sinners, alike, in order to appease Satan's appetite for lust and destruction against all souls. Yes, it is all mentioned in the Bible and in many books of literature that can be found in any large city library.

The cavern world is not as it is portrayed by movies, especially in the adventures of those who explored in "Journey to the Center of the Earth" - having the types of clothes that the movie portrayed with a lady wearing a corset and taking a pet duck on this journey, etc. The movie "SHE" was better and closer to the truth than "Journey to the Center of the Earth," which did give a good concept of cavern life.

ARE WE LOST?

The cavern world is not "lost," just misplaced in the superstitions of our forefathers and their fathers – fathers who cling to the LOST memory of their past. Our "heritage" is also lost because of the mistakes of our forefathers. They wanted a king, they got it; they wanted law, lawyers, politicians and generals, they got that, too. Worst of all, they wanted mumbo-jumbo, rituals and secret societies with long black robes. That, too, they got. Then, they went to whoring and pimping among the nations, creating harlots' amongst their unstable religious orders. They removed knowledge and created ignorance, and men turned into monsters.

Superstition and fear ran rampart at every turn and in every corner, and the "Black Robes" stood as though pious, saintly and godly, crying out against those they called witches. The ignorant called out, "BURN, WITCH, BURN." All these things, your forefathers allowed to happen. Now, the world stands at the end of its civilization. Because of all their witching and false prophets, they will have their witchery BURN.

Yes, we have LOST our heritage and what the Elder Gods left us.

IS THIS THE LAST FRONTIER?

This is not the "last frontier," but it is the last of the old way of life and the start of renewing the pioneer spirit which was left to us by our fathers. That same spirit is still in the minds of many, with the adventurous spirit for the many frontiers ahead. Still, in a sense, the cavern's secret world is the last frontier for men on (and in) this planet, Earth. But, there are new frontiers, yet to be seen, beyond anything which you can ever conceive in your mind, at this time, as mankind still has old superstitious ideas from their forefathers. Worse yet, mankind fears to give up old cherished ideas of our surface society, which is based on prejudice against all mankind.

Yes, I, George Wight, live in the cavern world, some fifteen miles or more below the earth's surface, with the secret people, called Teros, while you sleep, under your stars by night and are blistered by your sun by day. I gather knowledge from the Teros and records from the Ancients while I am completely surrounded with a constant green, soothing light which comes from everywhere. This light comes from the walls, the floors that I walk upon, and, yet, the light seems to be from nowhere. The caverns, farther below, are endless, and it is like being out in a boat in the middle of the Atlantic Ocean where all you see is sea.

There are many frontiers and horizons, yet, to be discovered. There is even a Shambhala where hidden kingdoms are bathed in a golden rule, ruled by none, in remote places deep below the surface, separated by other secret people. Then, there is yet to found, the center of the earth, the "Inner Earth," populated by an entire race of people, far beyond your ideas, and not according to the writing of Gardner, Raymond Bernard and Rev. Blessing. Also, there is another frontier, yet, to behold – to the "STARS." It is not through the "POLES," but beyond that, for the Gods are coming.

The Teros told Mr. Wight that they believed that the old Gods will come by the turn of this century to free them, and the Elder Gods win take them to their kingdom, to the stars and beyond. That is their belief – maybe, a prophecy – but it is believed by the people of the cavern, as they all tell the same story that the Elder Gods will come back and restore our rights and our heritage. I (Marcoux) have predicted many times that the UFO will come around 2003 AD and before 2007 AD and take many people away from this world. Will the "meek" inherit the Earth?

INTRODUCTION TO GEORGE D. WIGHT

In this issue, I am introducing you to an old friend whom I have known since 1956. In June of 1970, George D. Wight left a series of notes for me and for four other persons who shared this experience, some fifteen miles below the Earth's surface, and they met and lived with the secret people, whom you call "TEROS." George was a UFO buff, writing articles for several zines, mainly with a UFO magazine in Michigan, having his own followers. He was also a spelunker – not just an amateur, but was very active in the state of Missouri, especially in the Ozarks. He went in many caves and caverns with a group who explored in Missouri, Arkansas, Kentucky and Tennessee.

My association with George was more because of his open-mindedness about the Shaver/Marcoux hypothesis, as he called it. Even though we got along well, we did have different opinions about the cavern world, and he could not accept some of the Shaver/Marcoux concepts. George was also associated with one of the members of the UFO magazine, and they both knew about

Shaver, Palmer and Marcoux concepts, as both visited my home in Flint, Michigan, many times. They were able to publish a UFO zine, which I supported, giving both sides of the story, even though they didn't believe in the underworld, the cavern world and, especially, about Teros and Deros.

George gave his diary to one of his friends that had shared in this experience with the Teros, a school teacher whom I also knew. He promised George that he would get this material to me, if at all possible. It took twelve years for this old friend to get this material to me, as he tried many times to get in contact with me before succeeding. Strange as it may seem, I was in Michigan at the time of his experiences, just a few hundred miles away. I often thought about them and other associates I'd known from 1955 to 1960, but we lost contact with each other. This is just a brief background about George, and the others that shared this experience with the Teros do not want me to reveal their names nor where they live. There are only twelve people that know about this, besides me, but, now, you, also, can share this valuable information.

WE MEET WITH THE TEROS

George wanted me to receive this valuable material because he felt that he owed me something, since it was me who had introduced him to the Shaver/Marcoux hypothesis in lectures which I gave in Flint, Mich., in 1956-57. In his own words, "Charles, you're a great guy, and sincere, but in all my years as a spelunker (since 14 years old), I had never seen, heard nor believed in the Hidden Cavern World ideas which you and Shaver proposed."

Later, Wight says, "We explored some caverns in the Ozarks and investigated Bat Cave and some other known caverns, which are many in these mountains. In 1960, we did extensive exploration in this one particular cavern which constantly seemed to draw us to it. As I now look back, I was unconsciously, yet somewhat consciously, thinking of your words, since they just seemed to ring in my mind as we continued to explore deeper into this fantastic cavern. At times, I just couldn't shake your words out of my mind, but I kept this to myself. Of all the people that were with me in this adventure, no one knew anything about your ideas, except the one person who was in from the start of the UFO magazine in Mich. While alone, we would both just laugh it off about the Shaver, Palmer and Marcoux hypothesis."

FURTHER EXPLORATIONS

Between 1960 and '67, we would find time during our vacations and holidays to come back to this cavern to explore further, curiosity, or something, just drove us on. In the last week of Dec. 1966, and the first week of Jan., '67, the five remaining of-this original group, came to the end of a certain narrow tunnel. The tunnel was off from a chamber which we called Glass Cave, where our main "stake marker" was located. The tunnel is not too far from our marker and runs down deeper in the Earth, about a mile or more. As we got to the end of this tunnel, we could faintly see a greenish glowing light.

At first, we thought that our eyes were playing tricks on us after being in these dark caverns more than five days. After we left the Glass Cave, we were sure that being in darkness for so long with only our carbon light was causing this green glowing light before us. All of us saw the light, and all of us believed that darkness was directly affecting our eyes and making the carbon light look green.

Curiosity made two of us go toward the end of the tunnel and probe around to see if the light was possibly coming from some other source, as we were beginning to wonder about it. As we went toward the light, we noticed a "CREVICE" barely wide enough to get in. I crawled in several feet where I came upon "MAN-MADE STEPS" and I called the others to follow me and bring the equipment with them.

The opening became wide and high enough to walk upright. When we walked down the steps, the green light became more pronounced, even though we still had to use our carbon light. As we proceeded down these steps, we suddenly came into a large tunnel, or corridor, about twenty feet wide and just as high. The walls and the floor were smooth, and the ceiling had a curved, "dome" shape. We knew that this corridor was not a freak of nature, but MAN-MADE. As we all stood in this corridor, we could see and feel the bright, green light in the distance. It was darker in back of us, so we went toward that green light.

WE STUMBLED INTO THE HIDDEN WORLD

We had accidentally stumbled into the secret cavern world! All of us were appalled and dumbfounded, so to speak. But, the two of us did not mention Marcoux's concept, even then. Yet, we looked at each other and questioned, in our minds, "IS MARCOUX

RIGHT, AFTER ALL?" At the same time, there tumbled over in our minds, a helter-skelter of thousands of ideas, still mixed with doubts about the many things that Marcoux had said, and we didn't want to believe, yet, were afraid not to. Our mouths just hung open at the wonders of this fantastic corridor, Charles; your words seemed to ring like a bell, and I just knew that you were right. Yes, Charles, all that you have told us is true, and some of the places where you said there were entrances to the hidden cavern world, really do exist.

Now, that I live here, somewhere in these secret caverns, I can tell you, for certain, there are things which even you do not know. Maybe, you might know some of it, but there are things beyond your imagination. You can't know until you come here to see, feel and learn for yourself. I owe you a debt of gratitude, because the Teros healed my crippled leg, instantly.

I am grateful for more than just that, there are many other reasons, and I have left these notes and somewhat of a map, so that you, too, can enter and visit with these people, who some call Teros. I have no desire to come back up to the surface, so, maybe we will meet here, some day, where I now live with the Teros, in these caverns of the ancients.

It was the last week of Dec. 1965, when we suddenly became aware of a bright light seeping out of a small crevice from the floor of the tunnel which we were exploring. Not being sure if we were just seeing an unusual light or if it was caused by our carbon lanterns, we proceeded to investigate the source of that light. The crevice was just wide enough to crawl through, and since I was the smallest, I started lowering myself through the crevice. It was very narrow, with various obstructions, but I was able to make my way down, feet first. The others tried to guide me through with encouragement and cheers. It didn't give me any comfort to not be able to see what was below me as I slowly struggled my way down, step by step, with determination. I hoped that I would eventually come into a larger opening that would make it feasible to move about more freely. As I groped for footing, I suddenly lost my grip, somewhat, and my weight caused me to fall, tumbling downward.

I landed in a prone position, and I rolled over to recover my composure and examine myself for injuries. Finding only a few bruises, I looked toward the light in the distance. As I stood up and observed my surroundings, I called the others to join me, assuring them that I was all right. Surveying the corridor I found myself in, I was very aware that it was not natural, but man-made, and I was just appalled at this strange corridor. Then, I gave a hand to help the others through that crevice. Turning our attention to the light in the distance, we thought it was an exit out of the cave.

One of the men reasoned, according to his calculations, we were down at least five miles, maybe seven miles, at the most. He remarked that we couldn't possibly be on our way out to the earth's surface. After we went about ten miles further down that corridor, we discovered that the walls, themselves, were the source of that light. Don't ask how or why, for we did not know the answer. It was difficult to understand how it could be. It was, then, that I told the others the whole story of Marcoux, Shaver and the hypothesis of a cavern world. This gave them a new insight as to what they faced ahead, if they were to proceed. After much discussion, we all agreed to explore further, in hopes of finding the Teros and, eventually, to explore to the CENTER OF THE EARTH.

There are strange feelings when you get about a mile down into the earth, and the further down we went, the better we would feel. We got more energy and ate less, only twice a day, as we explored deeper into the Earth. Since there was no way to positively determine our distance, we only have a rough idea of how far we walked into that corridor. When we came to a place which we call the "ten mile point," we could see various types of animals mingling in the forest of the cavern world, beyond the glass-like walls of this corridor. Some of the animals appeared to be wild boars, and they were as large as a rhino.

At one point, we came upon a monster stomping and pounding at the wall, frantically trying to get at us. We weren't sure whether that "THING" could get through the wall, or not, but we had our weapons, and this gave us courage to hold our ground. The thing wasn't able to break through, and we learned, later, that it was what you call a "DERO." It was horrible and most frightening to encounter one so closely. The Dero was about eight to nine feet tall, bushy haired, brownish in color and with a face that no mother could love. Perhaps, you can visualize it as a combination of a boar, an ape and a mingling of unknown species from the long forgotten past.

WE MEET THE TEROS, THE SECRET PEOPLE

Suddenly and unexpectedly, after we passed one intersection, three tall, strange people stood before us, as if from out of nowhere. They were about nine feet tall, wearing strange clothes of strange design. Surprisingly, we were not afraid. It was as if there was a telepathic communication between them and us, giving a feeling of peace and comfort, with complete assurance of security. Their language was most strange. Also, upon their chest and wrist was some kind of device which enabled them to communicate with us. They led us to a door and into an ELEVATOR, which took us deeper into the earth, a distance of about fifteen miles (so far as we could guess) almost instantly. The "El" gave no sensation of going downward that comes from surface elevators.

As we got out of the "El," there was a city before us with a civilization beyond your conception. The cavern was of some form of glass, with perpetual rays of light and had a comfortable, even temperature. There was a feeling of peace and of physical energy, which gave some form of stimulation to our mind, awakening our mentality. The depth and size of the cavern can only be guessed at.

These secret people, which we call TEROS, have a society that is completely alien to our physical point of view. They cannot come up on the surface, for the SUNLIGHT would kill them in a few hours. They say that the SUN is the cause of death, for it speeds up the aging of life forms on the surface. They are able to live for centuries, due to the fact that they do not live near the sun's direct solar rays. The people are much the same as us, in many ways; they live and play very much the same as we do, but their approach is vastly different than ours.

Because of their beliefs, they can't think as surface people do, and, also, because their physical bodies are highly stimulated by the constant light that permeates their cavern world. The light, too, aids their ability to live for many centuries, and their Knowledge is multiplied by the additional time between their birth and their endeavors in life. It is quite apparent, from what the Teros told us, that this constant life force energy comes from the inner earth, itself.

Their homes and other buildings are carved out of the glass walls, itself, as well as their furniture. It is almost impossible to explain their various machines, their medical system or their society. They have no politics or religion, by surface views, for it appears that they seem to be guided in their beliefs by their one-mind idea of one for all and all for one, in that sense. The people are about seven to nine feet tall, with the men being taller than the women, with a few exceptions, like us.

Their faces are narrow and appear to be longer. Their skin is gray, due to lack of sunlight, not like surface people whose skin is reddish or brown, etc., due to solar radiation. Their eyes are very large and somewhat owlish.

The corridor that leads to the Tero's secret city.

Their bodies are firm and solid, and you can really sense their physical strength, which is beyond anything you can imagine. The women, too, have great physical strength, yet, are more delicate than men and their children have the same healthy appearance and are as active and playful as surface children.

Generally, their society has no bickering, fighting, murder or wars, but they have all the means for war, if it becomes necessary. On rare occasions, if one of their own does get out of hand and deliberately injures another, then, their council, so to speak, decides whether they can stay or are to be cast out to wander, alone, in the many corridors and caverns o they are vast beyond your concept. Their population remains at an almost constant level, so far as I can determine. It's as if they have a means to limit their population by the power of their minds. Their beliefs are like a science of minds with all minds being one, and they make it a form of religion. They religiously practice their beliefs as a way of life in regard to their freedom, their consciousness and their gentle approach to all life. Despite their great size and strength, the secret people are gentle, kind, considerate, compassionate and extremely friendly.

We were allowed to go deeper into the earth by means of an elevator, and there are caverns so vast that you can't see the end of them. Some are so void of life that the echo of your walking and the sound of your voice is all you hear. Many of the remains of the ancient cities are still there, intact, though. In many cases, their buildings are sealed by some system of the ancients. Some others are still open, and you are able to enter and investigate the many wonders of the gods of the past. The Teros will allow you to explore and investigate, but they will not allow you to use, or pilfer, any antiques for your own use. They constantly keep guards on duty that use ray mechs to search the many caverns and corridors.

The four of us who explored this cavern were two teachers, a medical technician and a deputy sheriff. It was because of our desire to help and heal people that the Teros accepted us into their cavern and showed us some of their sciences and many wonder machines. One was a highly sophisticated machine which speeds up healing and is able to prolong life much further than normal. In their second trip to the cavern world, one of the four took multiple samples of skin and fingernails from his patients, and the Teros analyzed them, giving their findings on how to help heal them. But, the medical profession refused to listen to the information, and another piece of knowledge was lost because of the stupidity of the medical profession.

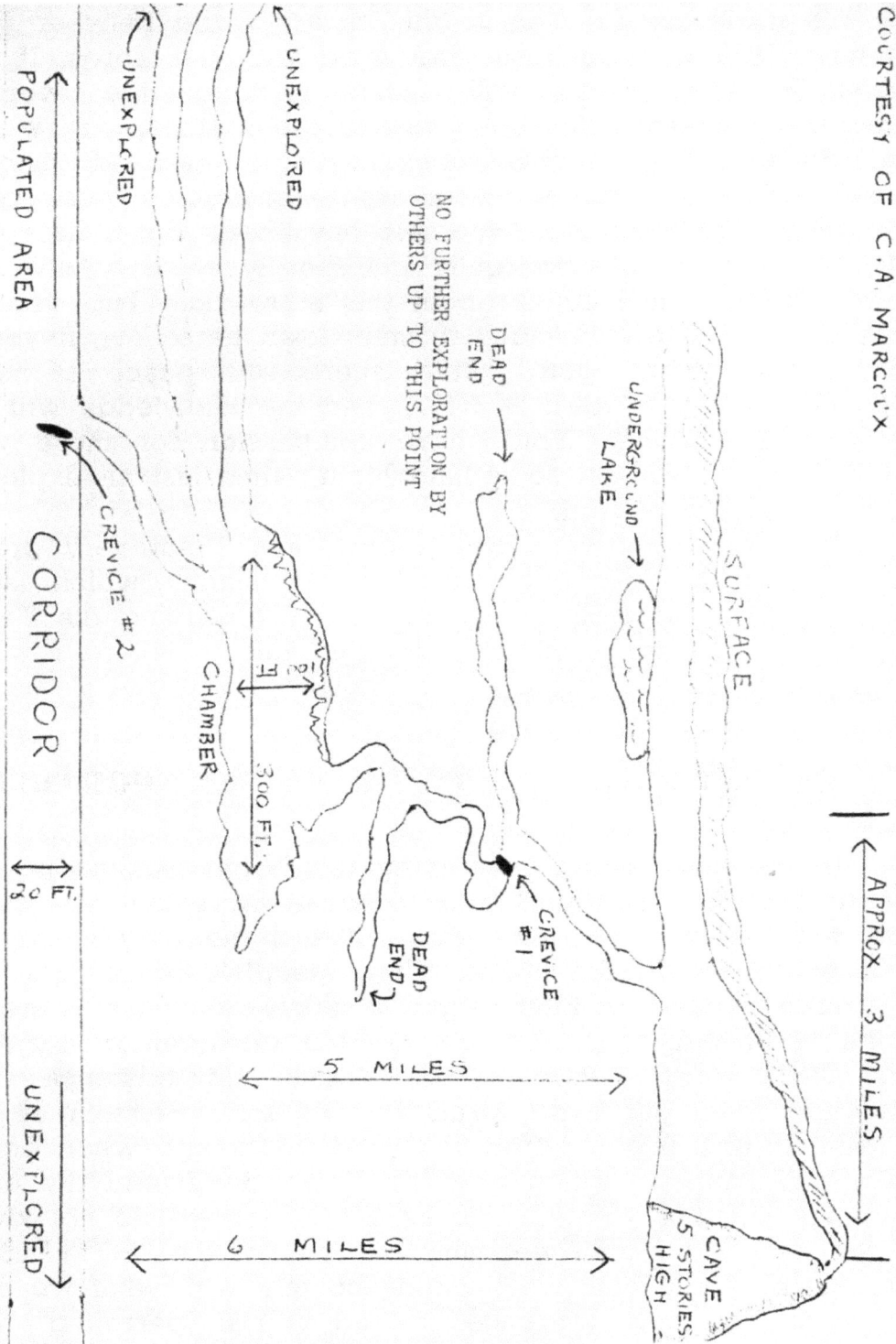

G.D.W MAP 1967-70
COURTESY OF C.A. MARCOUX

SURFACE

UNDERGROUND LAKE

APPROX 3 MILES

CAVE 5 STORIES HIGH

DEAD END

NO FURTHER EXPLORATION BY OTHERS UP TO THIS POINT

CREVICE #1

DEAD END

UNEXPLORED

CHAMBER

300 FT.

20 FT.

5 MILES

6 MILES

UNEXPLORED

POPULATED AREA

CREVICE #2

CORRIDOR

UNEXPLORED

NOTE FROM CHARLES A. MARCOUX: In this issue of The Hollow Hassle I have dealt with a cavern entrance into the secret hidden world, but I will not reveal its location at this time. I am planning an expedition into this cavern, where we can meet the Teros, and I will offer the readers an opportunity to join this expedition. The story is true, so far as I know, and I feel it's worth the gamble.

53

We made several trips to this cavern world between Jan. 1966 and June, 1970, our last visit, and lived with the Teros. If you try, Charles, I predict that you will succeed in finding the cavern where I now live. When you return back to the surface, you will be so frustrated with the misconceptions, lies and unrelated things about the subworld that you'll not even try to change anyone's views on the surface. Only after you, yourself, have been down here will you realize what the subworld really is. You'll see how futile it is to try to tell what the truth is about this subworld. But, who knows? You may be one who decides to stay down here. Yet, if you do go back to the surface, you'll return there with peace of mind and conscience, because you'll "KNOW," and the knowledge will set you free from ignorance. You'll have compassion for those who are misled, and you will be something higher than just an explorer, you will be a teacher of truth.

The Teros are wonderful people, and I am very grateful to them for healing my crippled leg, almost instantly, which astonished the other three, as well as myself. After observing the Teros and their way of life for a while, I decided this is where I wanted to remain for the rest of my life, so, I now live with the Teros.

SECRET PATHWAYS TO THE ANCIENT PORTALS

In this next section concerning the Wight manuscript, I intend to give you another "first" and to clear up some misconceptions about the Hollow Earth, the Subworld and the people who live in the ancient cavern Elder World. Those people's ancestry is lost, and the source of their genetic origin is unknown. They were in and on the earth and, of course, beyond the reach of gravity from the stars to the earth. Perhaps, they came into existence a million years ago when our earth formed its extensive system of caverns. At least, it was one hundred thousand years ago when Lemurians, Azaltans (Atlantians), Titans and the Norse dominated the earth. It is not my intention to state this as a fact, for we do not know, nor will we ever know until we enter into one of the "PORTALS" to the inner earth. Until then, we must be content with what we do know and search for more evidence, instead of just "parroting" what has already been said and being content with that. I will never be content, nor should you, until we actually get into the subworld and meet with the secret people, known as the Teros. Neither should we accept things as being hopeless. I do not accept anything as being hopeless or become hypnotized by the mind manipulators.

THE BEST OF THE HOLLOW HASSLE

This chapter will deal with the Wight manuscript and also with my own personal experiences of the subworld since my first "vision" (in 1920) and of my search and investigation of this hollow earth, which is often referred to as the "Inner Earth." I'll present information from my own personal manuscripts pertaining to the inner earth and related subjects concerned with the subworld and the people that live in the ancient Elder World. The manuscripts were written and illustrated in 1955-56 when I first met Wight and two others that were with the "four." The result is a combination of the Wight manuscript, material received from the "four" and my many manuscripts, which have not been published, as yet.

VENUS, A HOLLOW PLANET

The Russians, for some uncertain reason, have sent twenty-seven probes to the planet, Venus. Our own government has sent out several probes to investigate this same planet. The magazine, Science Digest, recently showed the planet, Venus, on the front cover. Inside, it also showed that Venus was hollow with several tiers, or levels, within it and an "opening" at each of the poles. It shows a circular opening on some lower levels within this hollow planet, Venus. It told that "within" the center of this hollow planet is contained a combination of gases. Therefore, life in our present form, as we know it, makes it impossible to exist on, or within, Venus. The article says that the "gas" is similar to methylchloride. If this is true, and their probes seem to prove this to be so, then, the inner configuration of Venus must be very cold. Methylchloride is a poisonous gas that is used for refrigeration and is in organic synthesis. Further, the various probes to Venus seem to imply that Venus' surface is extremely hot, but within the hollow center of Venus, it is very cold. Therefore, there has to be some very extreme changes going on in the physical structure of Venus. The two forces, "hot and cold," must clash and repel each other.

In other words, the surface has some extreme heat that creates currents above its surface, as well as within the planet. The "gas" which is contained within Venus must exit from the "poles" to the surface, and vice versa, causing extreme currents around and within it to become a vicious circulation of the cold gas as it meets the surface extreme heat. This would create cloud formations above the planet surface with a high reflection effect, and this is why Venus is called the "Bright and morning star," as referred to in the Bible ("Why art thou fallen, oh Lucifer, thy bright and morning star.")

OUR PLANET, EARTH, IS HOLLOW

For over a century, give or take, our world's scientists have claimed that the earth is a geode shape, earthlike object. A "geode" is hollow, with a cavity. Now, after all the probes to Venus found that it is hollow, I wonder just how many probes on our own planet, earth, have shown that it is also hollow, with openings, or "portals," at the poles. Why have the scientists shown that Venus is hollow but refuse to tell us of the true configuration of our own planet?

Why do our scientists continue to tell us that the center of our planet is a hot, solid ball of iron, surrounded with a liquid, or lava, that gives our earth its heat? Then, the very same scientists claim that our earth gets its heat from the sun's rays, and not from the earth, itself. I find these two claims are not compatible. Now, both cannot be doing the same thing. The so-called ball of iron is "shielded" by hundreds of miles of the earth's solid crust, and the sun cannot penetrate its heat and life to the very center of our earth's so-called ball of iron. These are questions that you should consider and which I intend to clear up.

Scientists and intelligent individuals claim that as one goes deeper into the earth, the temperature goes up and, thus, no one can live 5, 10 or 20 miles down into the inner earth. But, they fail to explain why deep oil and water wells give up oil and water with temperatures in the 50 degree range. NO ONE has explained why subterranean caverns have a stable temperature, regardless of the depth. Furthermore, the scientists and professional experts have yet to explain why some caves have air exiting from the inner earth that is cold and refreshing with its temperature in the range of 58 degrees, give or take. If the inner earth is hot, then, why do ALL BLOWING CAVES have cold air? It should be very hot air, like coming from a furnace that could be harnessed and put to good use for us.

Spelunkers believe that all air that comes out of blowing caves has another "exit" from which the air is circulating. They believe there must be some small cave or crevice which draws the air from the surface, but they are never able to find its source. They do not believe the inner earth is hollow, since they cannot think in these terms and cannot rationalize that such a thing is possible. After all, the professional experts have brainwashed them to follow the thinking, or the "image," of the learned professors and scientists, for the spelunkers and scientists have claimed that there must be an exit for carbon dioxide to be drawn out. Not so! At the depth of seven, or more, miles down into the earth, the plant life is extensive. George Wight says the trees never shed their leaves, unlike trees on the surface. Also, the trees and other greenery are huge in size with very large leaves. The plants and trees absorb carbon dioxide and give off oxygen. As all the plants and trees are exposed to beneficial rays, I believe this causes their huge size, but I can only guess.

NOTES AND QUOTES FROM WIGHT MANUSCRIPT

In the first section of "I Live With the Teros," Wight and the four said, "If we had a car, we could travel to the very center of the earth." So far as we know, that is about 3,410 miles. They mentioned that they did explore some 70, or more, miles into the subworld, as well as the forests outside of the corridors. In the seven years of their exploration, they recorded many experiences which they refuse to tell because of being rejected by those who "know-it-all". "In the beginning, when we first entered the other world, we told everyone - including reporters. They stuck the story with other UFO stories. Some even attributed it to a mystical experience, psychic phenomena, hallucinations, but most just didn't believe it. We asked two reporters to go with us. We bought their gear and equipment and paid for everything. At the last moment, they declined. Even our many phone calls and letters to Ray Palmer were rejected and nothing was mentioned in any of his publications, and Shaver never replied to our letters. The Teros were prophetic when they told us that their 'greatest strength' was in the fact that THEY know we exist, but WE do not know THEY exist."

The corridors are about twenty feet across and just as high. "When I first fell into the corridor, I really believed I fell into an old mine except "J", who always had perspective, stated that there are no mines in N. America at a depth of about seven miles. Also, the entire corridor resembled CERAMIC, with crystal clear see-through walls. Cushman has many old manganese mines. When I was in high school, my brother-in-law and I mined manganese for three weeks, and this is why I felt that I had fallen into an old mine."

I found that the manuscript from George Wight left me with hundreds of questions to ask, but I have to depend on first hand information from one of the "four" whom I personally met in 1955. Sometimes, information is sent to me which I haven't asked, too. Twelve persons originally began this venture, but only four, along with Wight, were able to do any exploration into the subworld. All five were single and had no commitments, so this made it easier for them to explore the subworld and "live with the Teros." I asked the spokesman for certain information and he said, "As I sit here, I seem to think back and wonder why I didn't ask about a million things from the Teros. I guess I was overcome by the discovery, itself. I did realize one thing – the nocturnal animals on the surface descended from subworld animals. After thousands of years, they have diminished in size and capabilities.

There are bats down there in the dark, next to the corridors, with wing spans of five and six feet. They are ugly and vicious things. "J" brought a moth, with a wing span of 20 inches, to the surface.

The Teros caught it for us as a souvenir. We arrived back on the surface on a June day. "J" opened the container to view it in the sunlight, and the moth began to turn brown, then black, and it charred to ashes. If any similar thing ventured to the surface, I don't believe they'd be able to survive unless they come out only at night. Also, Teros are not able to exist on the surface, for their environment has not changed for thousands of years."

Long before the four gave me the manuscript, it was always my opinion that if animals from the subworld came up on the surface to the ultraviolet rays of the sun, they would soon wilt and die. Many animals have been seen at night which has been described as monsters, such as Big Foot. Around Cushman areas, some people don't venture out at night, after seeing and hearing unusual animals. Some have even seen what they call "CAT MEN."

In 1952, a circus was to be held in Batesville, Arkansas. When the circus train was just about to enter the city limits of Batesville, it was derailed, and many of the animals were set free and scurried off into the mountains surrounding that city. Some of you may remember this situation which was featured in most newspapers in the USA – I believe it was Ringling Brothers. They claimed that they had captured all the escaped animals, but many people around Batesville had doubts that all were captured. The reason is because many natives have seen unusual animals and heard animal sounds in the mountains surrounding Batesville and Cushman. There has been a story in this area for years that many have seen what they call a "cat-shaped man that snarls just like a cat.

A co-worker friend of mine, who worked with me at General Electric in the early sixties, was born in the Ozarks, and he said that he used to explore caves. One day, while he and several buddies were exploring near the river between Kansas City, Kansas, and Kansas City, Missouri, they stood on a high hill and could see a very large cave across the river. So, they crossed the river to investigate it. Not having any ropes with them, they found some loose wire fence which they secured, and two of them climbed down to a ledge about 25 ft. below the top of the cave where they were able to swing around into the cave. Now, Duane had his father's old double barreled, sawed-off shotgun, since there are many bears and Razorbacks in these mountains. They walked into the cave about 25 or 30 ft. when, suddenly, a huge animal jumped out at them from the darkness, striking one guy on his right shoulder and knocked him to the ground with a very bad wound. Immediately, Duane shot both barrels of his gun right into the beast. It knocked him down and gave them time to climb back up their wire-made ladder to the top of that cliff. The beast was "something from hell," as he called it, and it snarled like a cat and tried to get them as they scrambled up to the top. He said he will never go in another cave again. As he described it, the beast was about eight feet tall and walked like a man but looked like a tiger with huge teeth.

Wight and the four say they saw very large saber-toothed-tigers in the forest beyond the walls of the corridors. Many unusual animals that live in the subworld do often come up on the surface, like the large Black Panther seen recently in Pennsylvania, and some have been recorded in other parts of the USA. The Loch Ness monsters come from the subworld, and, in some way, have found their way to the surface, just like in many other places throughout the world. Most UFO experts claim that monsters were brought here from another planet by UFOs. This is only their guess, and they confuse the facts, for they do not believe in the existence of the Hollow Earth or life in it. I want to remind everyone to never go into caves or caverns alone, and never go in UNARMED – be prepared.

LIVE FOREVER

Many people have the desire of the ages, to live forever. I know that we do live forever in the sense of the mind, only, or what is called the "soul," for a better explanation of eternal life. The mechanism, called man, can achieve a certain level of immortality in the physical self, but the physical self must separate so that the mind can explore eternal life and reach true immortality. The mind never dies, that I know, but all else seems uncertain in the flesh.

In speaking of a long extensive life, such as Adam, Eve and others mentioned in Genesis, the Teros, too, have a limit and do die. The average age in their world is from three to four hundred years. They, too, do not have the answer to a long extensive life such as Adam and Eve, but Wight and the four were shown the reason for the short life span on the surface, and the cause of aging is simple. In time, science will catch up and verify my statements as to its cause, which was shown by projection to Wight and the four.

EXPLAINING DNA – ANOTHER FIRST

DNA is the body's master molecule. It determines what we are and how we function. The DMA molecule makes up genes. Life is controlled genetically by a mechanism within the DNA. But, on the earth's surface, the DNA is constantly being bombarded by ULTRAVIOLET LIGHT FROM THE SUN which causes damage to the DNA. Normally, enzymes repair the damage. However, after several decades of exposure to the ULTRAVIOLET LIGHT, the DNA is damaged beyond repair. Then, the damaged DNA sends an incorrect message to each cell, causing it to break down and cease activity that causes a loss in the normal life processes.

In a nutshell, this is what causes aging and deterioration. ULTRAVIOLET RADIATION affects every cell in one's body, and since cells make up every organ, gland, tissue and bone, the effect is total. It may be that if one spends time in the subworld, away from ULTRAVIOLET LIGHT, one will begin to feel physically better, because the DNA will not be further damaged and enzymes will not be overburdened with repairing cells. This is also the reason the Teros (Deros, too) and animal life have great size. Their cells have a chance to grow without the continuous destruction.

GEORGE D. WIGHT, PAST AND PRESENT

Wight was strictly a "UFO BUFF" and not a believer of Shaver, Palmer and my ideas of the subworld. His interest in the subject was limited to the nature and geology of caverns, only. All else about my claims of the subworld and the white rays, located in Arkansas, Missouri and Kentucky, just went over his head, for his mind was completely centered on saucers.

There was a very large UFO group in Detroit that published their own zine, a group of followers in Flint and a UFO club with a zine in Saginaw. Wight made a trip to Michigan to visit with two of the "four" and myself, and, later, was to make a trip to Canada to visit with Gene Duplantier, publisher of *Saucer, Space and Science*, who all of us corresponded with, once. Also, Wight had his own zine, at the time, and was making a trip to the well known saucer clubs in Michigan and Canada. His plans were to meet with Donna Heathcote and, then, to meet Duplanteir and go back to his home town of Bedford, Ohio, through Wisconsin.

We, and the "four," never gave it a thought, nor asked him of the results of his trip to Canada to meet with Heathcote and Duplantier. But, now, as I think about it, and after asking one of the "four" about his trip, we find that none of us asked him about it. With the knowledge of Wight living with the Teros, it would give us a form of "PROOF" that Wight does, and did, really exist. He did live in Bedford, Ohio, and published his own saucer zine.

George was a very likable gentleman. He was thin, about six feet tall, even standing with his leg braces. He never used profanity, and even though it appeared he was helpless, he was always eager and happy to assist others. "He was in excellent health, and walking, when we last saw him (1972), and even with more freedom of movement, he was still a pious and gentle man."

Regarding proof of Wight's existence, the "four" could verify his existence, but it would expose the "four," which I will not reveal. This would mean that the four would have to prove the PLACE of George's existence.

No one has come to believe in the existence of a subworld, and there lies the crux of the proof but also the security of George's existence. The Teros were clever in their endeavor. First of all, they predicted that in order to prove George was "down there," it would be necessary to prove the existence of a subworld. I had a friend who majored in computers and made a check on George's Social Security number, driver's license and address. As far as she was concerned, these numbers never existed. Even George's street doesn't exist. George didn't leave a family behind. Proof of George's existence on the surface has vanished.

The secret four and their club members never accomplished their desire to travel to unknown depths or follow the corridors to wherever they led. On the final trip (12/72) they made, they learned a lot from George. "The subworld is endless and spacious, and one cannot possibly see the entire area because of the various layer upon layer of caverns. There is also a part of the subworld which is not used for any purpose except as part of a corridor between populated areas. In these areas, the corridor is enclosed in glass-like material (like a tunnel) to permit traveling from place to place without disturbing the environment and, also, to keep out animal life."

"There are ancient caverns which used to be large enough to permit animals from the subworld to wander to the surface. In some instances, several species which traveled in herds emerged in Siberia. However, they couldn't survive. First of all, they were photosensitive and the sunlight destroyed their vision. They were helpless, and the whole herd perished in one huge pile. This herd resembled elephants and are called "mammoths." I'm waiting for the Russians to claim they found a herd of prehistoric elephants, or their remains.

"These cavern entrances were destroyed by the Teros so that no others could come up on the surface to be found by some nosey Russian scientist or military personnel who might suspect the true source of the elephants. Recently, an article in *Science Magazine* stated that African natives, in the bush, reported seeing huge beasts which, according to descriptions obtained by researchers, resemble dinosaurs. Some natives claimed to have killed one, for it was destroying their fish sources. In the subworld, animals also travel in groups, and if several find their way out, it's only a matter of time before the entrance is sealed."

"I have always had a dislike for spiders and snakes. My fears diminished to some extent after I, and the rest of the group, viewed through the corridor walls, spiders as large as truck tires and snakes as long as one hundred feet and thick as a bathtub. It seems animals have their own territory. However, if the Teros want to expand, build or repair, they have absolute control.

"To ease our fear and also to demonstrate the capability of their weapons, a Tero fired his weapon right through the corridor wall, but the wall remained intact. The end of the beam of light touched the serpent stretched out in the cavern room outside of the corridor for just a fraction of a second, but the whole serpent seemed to glow and nothing was left."

CONCLUSIONS

It is over 12 years since Wight first entered into the subworld and decided to remain and spend the rest of his life with the Teros. Since his manuscript made a promise to help me get into the subworld with him, there is no reason for me to believe otherwise and that he will assist in my endeavors into his world. His offer is also made to those that believe in the subworld, the existence of the Teros and decide to join forces with me in this great venture into the last frontier.

Since it is over 12 years, Wight may have explored into other avenues of the subworld, and his exact whereabouts may be beyond the reach of my thoughts to contact him. This does not affect my exploration to meet the Teros, for I am aware that the Teros have the means to scan my mind and will know that I am a friend of George D. Wight and the other four, whose names I know.

This will open the means for me and those that join forces with me. There are five "KEYS" to enter into the subworld, and the main key is Wight and myself. No others will be able to succeed without this information and the other four keys. Even I have to locate the proper tunnel that leads to the corridor, and it does concern extreme physical work, like climbing a mountain without proper shoes. But, in this case, we have to climb, the same as one would have to climb down a mountain. It is easier to climb up than to climb down.

This section was laid out and completed on April 22nd and mailed to Mary Davis, for the reason that I just may not be able to be around to do so after this is mailed. As you read this section, you will find that the last two sections contain a massive amount of information about the subworld that no one has, as yet, been able to tell (the real facts) about the hollow earth, including Shaver and Palmer. This is another "first," whether anyone believes it or not. It is your right to accept or reject it, according to your own understanding. This does not interfere with my plans and projects to enter into the inner hollow earth and, finally, to also "Live With The Teros."

The surface world is so full of "uncertainties" that after my many years on this surface, I find no reason why I should continue to live where crime is increasing and civilization is crumbling. Some of the subscribers of *Hollow Hassle* and *Shavertron*, as well as some of those that joined my Southwest Expeditionary Club, do not believe the story and my claims of the subworld and Wight. That is their right, but it is not their right to insult and degrade me, in order to gain attention from their own followers.

I am really disappointed with some, of all three groups, when they should be studying the information and searching out the fact and proof. I can't understand the fact that some spend so much money for zines and books on the Hollow Earth and claim to be a "believer" when their claims aren't compatible with their destructive attitudes. Yes, I am really disappointed. If they would just take the time to search for truth and knowledge, it could save their miserable wretched lives.

Those that join forces with me, and those that may follow behind us, will find me willing to assist you to enter wherever I may be in the subworld. Find Blowing Cave's south portal entrance and travel about one mile north into the cave; then, turn around and look back. If you can't see the light coming from the mouth of the cave, then, you know that you are very near the tunnel that leads to the corridor and the Teros. Meditate on me, Wight and the Teros, and they will scan you to see if you are sincere or not, If you are sincere and not "negative," I, or the Teros, will let you know and accept you. Remember, no one has ever made you this offer before, regardless of all the books that you have read with all their wild claims, including some that state that they live, or were born, in the cavern world.

Good hunting and good luck.

CHAPTER SIX
Straight Talk By Charles A. Marcoux
VOL. 4 NO. 2 – Aug. 1983

I have the book *Etidorhpa*, copyrighted in 1901 and in perfect condition, presented to me by Ralph M. Holland. Etidorhpa, spelled in reverse is "Aphrodite," as if the author was covering up its real meaning, perhaps. At any rate, it is interesting to note that *Etidorhpa* tells of the ancient Greek goddess, Aphrodite, who was love and beauty. Also, the MSS from Wight, and the other information I received from "The Secret Four" tells of a place of strange and foreign beauty, impossible to describe.

All the magical forms of their structures and designs seem to echo into one's mind, as if it was trying to tell us of their silent and shadowy presence. Like a ghost, it leaps into your mind, and the more you try, the more your mind falls far short of understanding. The secret four tell of vast forests and many forms of plant life, similar to that mentioned in *Etidorhpa*, in some degree, such as large fungus, mushrooms and other edible food which the Teros use for supplies.

In this cavern world, there still exist many forms of animal life that found their way into the hidden cavern world. Or is it the other way around? Many of the animals date from as far back as before the "flood" when Adam lived in the "cave of treasures," as mentioned in the "Lost Books of the Bible," such as dinosaurs, Loch Ness monsters and Big Foot that still lurk in the deeper parts of the forests in the subworld beyond the security of the corridors.

The "secret four" tell of their encounters while exploring through the corridors of the subworld, and in their very own words, "The four of us had no responsibilities or someone to come back to. We used to talk of following corridors to other parts of the interior of the Earth. Around the ten mile depth, we were shown life and dangers outside of the corridor, through transparent walls. Did you ever wonder where rats come from when they sometimes swarmed out of the ground, or did you ever see ants as big as your thumb? We saw wild boars as big as rhinos, roaming in the caverns outside of the corridor. We saw saber-tooth tigers on two occasions, and there are probably other things roaming in the dark corridors that defy all explanation. There is nothing in one's nightmare to compare with a Dero who's mad because he can't get at you, and he continues to pound the corridor wall, but to no avail."

Of the books compiled by Bruce Walton, the front cover of Volume Two shows the cavern world with a maze of "glass tunnels" going in various directions. That is quite similar to the cavern world described by Wight, especially about the glass corridor. Although the drawing is far short of the actuality of the subworld and the Teros, for this society is isolated from the outside cavern world. But, they are able to get into the forest for necessary substances, as well as for the animals which are similar to cattle. Yes, they do eat meat, but the difference is that they know how to cook meat to eliminate "disintegrate" elements from their food.

Because of the ways they cook their food, the surface people are riddled with various diseases. Teros also tell us that many children are born mentally and physically deficient because of the way the surface parents eat. Their own infections are passed into the fetus, which can make them mentally sick or deformed. This is also why some children who are seemingly born healthy are suddenly stricken.

People wonder why God has done this to their children, when it really started from their own eating habits, or came from unstable mental conditions which they, themselves, inherited from their parents, etc.

Have you ever seen ants as large as a man's thumb, or any other strange living form? In the Superstition Mts., near Apache Junction, Arizona, I actually saw a centipede that was a foot long and nearly as wide as my hand. Once, I also saw a beautiful light blue colored snake, over 20 feet long, which had just been hit by a car. It did not appear to be poisonous, and some of us examined it, but nobody knew what kind it was or why it was so large.

Such things, even, as our surface mushrooms come from the cavern world, and the reason some seeds get to the surface is because they are washed up from the underground water systems of the subworld. Yes, there are many strange things in the mountains, and in the many hidden caves and caverns that these very unusual animals and insects come from. So, there are still new frontiers to be discovered!

SHAVERISM AND CONFUSION

Shaverism was born in the January, 1945, issue of Amazing magazine, and along with it came Palmer's idea to make it a "MYSTERY." The word, mystery, has stuck ever since, causing more confusion and setting up a storm from new writers and false followers. There is no mystery, only confusion. Richard Shaver was about 70% right, and in my personal opinion, he was a great person, battling the odds against all negative forces, in, on and above the earth. You will find that RSS stated, in issue #1 of the *Shaver Mystery Club magazine* (1948-50), that he didn't know about anything beyond five or six miles below the earth. He suspected that there were deeper caverns, but he didn't know. Now, we know, for Wight's MSS shows, beyond a doubt, that there caverns all over the world and extending to the very core of this earth.

The cavern life, as portrayed by RSS and improved upon by RAP, was about the negative side of life, so, the present day readers and searchers are confused and full of fear about the cavern world. As you explore deeper into the earth, you get a lightness and more energy. The closer you get to the five or seven mile level, a feeling of freedom comes from the changing of the earth's gravity. Most of the caverns have no negative element, but there are isolated places where Dero-minded people live within five miles of the earth's surface, and that is what RSS encountered.

We have seen much confusion since 1950, after Shaver stopped writing about the secret cavern world. After 1950, Palmer never presented anything really "NEW" on the cavern world and neither did Shaver. Today, we have a new generation who followed misleading ideas, as presented by such as Adamski, Angellique and the followers of Giant Rock, Ca. This left just about everybody confused by the many wild claims at that time. Many hundreds of books on the subject were, and still are, confusing. The present day generation of searchers on the subject of the subworld and the inner earth are doing nothing but rereading and rehashing the material put out before 1950. All of them told of a one-person experience with no artifacts or proof, whatsoever. When confusion runs rampant, it utterly destroys your chances of getting into the cavern world. I have constantly warned people against the wolves who "sheer the sheep," which keeps you from getting into the inner earth. Yes, DEROS don't want you to know. They use people that are con artists, in their minds, with ego and other problems, so, this makes the "claimer" a "preacher of darkness" – you can recognize them as MIB (men-in-black).

THE BEST OF THE HOLLOW HASSLE

Many researchers of the subworld speak of it in the same manner as talking about the weather. In many cases, their ideas are just as changeable. They discuss the weather or use their library on subworld material in the same way that they rehash a sports event. The researchers have, yet, to really find much material about these secret people. Authors seem to neglect telling much about these secret people and their way of life.

They do not seem to get any meat in their stories, leaving readers confused with complex thoughts. Some people are dead-hard follow-the-leaders and can't be changed, or swayed, from the "image" presented by "their" authors. Yet, most authors gleaned their information from the writings of other authors, continuing to repeat the same mistaken ideas, but in their own way. Confusion is the best means to "PROTECT" the sub-world from the prying eyes of surface people, and it is the reason why the subworld (UFO, too) is never found, and will not be found, until we explore into the hidden world.

I will quote from a recent letter from one of the four: "Everyone talks about the Subworld, but no one does anything about it. If an entrance manifested itself in their presence, they would ignore it. You may have difficulty organizing a group. Everyone would rather sit home and read, or publish, misconceptions than finding the truth. You'll also realize how futile it is to try and change the minds on the surface and to try and advise on what truth really is. Shaver and Palmer never gave anyone a path to follow." There was only a path of misunderstanding of the true nature of the cavern world. This gave the followers confusion and fear of the cavern life, with ideas of the occult, bottomless pit and Satan-like Deros. But now you have the opportunity to enter into the subworld, without the fears and misconceptions, and find out first-hand facts and truth about the subworld for yourself. Who will be the first to set aside the confusion and misconceptions of past mistaken ideas and, first-hand, explore the world of the ancients with me?

By the time you read this issue, you will find that we will have made our first casual exploration trip to this cavern. In the Jan. 1983 issue of "Shavertron" and the Feb. 1983 issue of *Hollow Hassle*, I will present my findings. I have no doubt that I will find evidence of the hidden world and some photographs as "proof" that this cavern is the remains of an ancient "PORTAL" to the subworld, as an "evidence" of the remains of some of the antique artifacts when the ancients removed most of the pillars from around the opening of that cave. I will also bring proof of man-made steps that go down into the inner earth. Without a doubt, once you, personally, come there with me and see the remains of this ancient portal, you will never have doubts again about the subworld or of this ancient portal.

The secret four have recently given me more information about this cave. It was not from Geo. Wight, who omitted certain details that are necessary to further explore the subworld. They told of the opening of this cave being a portal (entrance) to the subworld. All around the opening of this cave entrance is some remains that are definitely man-made. In his own words, "The cave was once a portal, but it was changed, dismantled, and the blocks of granite, quartz and other material was removed and taken down below. Someday, you'll see the blocks all neatly stacked near the corridors. The portals and columns were removed three hundred years before the white man came on the North American shores. The entrance was allowed to weather and decay to give it the appearance of 'just another cavern."

Mind you, you will know that Wight and the secret four are telling the truth. They were the "FIRST" to enter and meet the Teros since the 13th century.

I have left something to think about for those that believe. This is for them, but not for unbelievers. Neither will I be argumentative with any of the readers of this magazine, or any other. Since the Aug. '82 issue of this article, I have gotten many unpleasant letters, and I will not waste my time on their views. Rather, I throw them in the trash. Also, there are investigators researchers, interviewers and some weak believers who seem to have an "AXE" to grind. Some are trying to use my material, to be twisted by their own views. Some try to bully me into their way of thinking and to be an image of their views - they think they know it all, when they do not. Neither will I be swayed by their intimidations or bullied into their way of thinking.

Due to these situations, I have copyrighted all of my writings pertaining to the Subworld, Wight and related subjects. No one can publish these writings, whatsoever, without my permission. Furthermore, the MSS and some of my select photography has been placed in a safety deposit box, and this is to let others know that I have taken steps to protect them and insure their safety.

Straight Talk
VOL. 4 NO. 2-Feb. 1983

Treasure comes in many forms, and the rich and wise put their treasure in banks for interest, while the man who buries his gold in the ground, or his Knowledge underneath a basket, receives no increase from the interest which he should have received. Treasure should gain interest so it can be reinvested in further wealth. The wise man that uses his treasures of knowledge by giving to mankind will receive a hundred fold in return of the treasures of his inner self. That knowledge is what you have received from the treasures of the heavens, where the mind intelligence was wisely put for the benefit of all mankind while still on this earth. Therefore, with all the knowledge that I have gained through the treasures of my mind and from my continuous endeavor to know of the unknown and to seek what I suspect or believe, I still strive to find answers to that which I know exists.

This is the third issue pertaining to the story of "I LIVE WITH THE TEROS." Now, I will reveal the location of this "PORTAL" or entrance to the subworld and the land of the "Ancient Immortals." With this article, also, I'm releasing all my responsibilities to all who read of this true story of the experiences of the four men that actually lived with the Teros. For the first time, people have a chance to see for themselves. Neither Richard Shaver nor Ray Palmer gave an opening to find a means to get to the subworld. So, remember, it is I, not Shaver or Palmer that give you readers an opportunity to find an opening into the subworld, meet the Teros, and discover the truth for yourselves. Any man who has great treasures of knowledge should not put it under a basket so that no man can enjoy the fruits of your work. Thus, this third article releases me of all further responsibility to all mankind, for it is now up to them to follow the information on how and where to get into the subworld.

SECRET PATHWAYS TO THE ANCIENT IMMORTALS

Just giving you the information and location of this Portal doesn't mean that you will find the means to get into the corridors and meet with the Teros. It does mean that only those who are sincere, without ideas to con the Teros for their own selfish welfare or do harm to anybody, whether they are your own fellow man or those that live in the subworld, can get in. You have to change your entire concept of the surface world and the people you have associated with in your lifetime. From the very words of the Teros and the manuscript from the four, they say, "Anyone who seeks to find and exploit the knowledge will receive no results. This is proven by the very fact that only a handful of individuals have actually entered the cavern world since before the coining of

Columbus. It is also easier to enter the subworld than to leave it. Their technology supersedes ours in finding out our true motives for being there. Our lie detectors are toys to them. The subworld people win not harm, but losing the knowledge of having been there is punishment enough." Let it be understood, here and now, that the knowledge of the subworld belongs to all who strive for TRUTH and the BENEFIT of others.

From Oct. 5 to the 13th, we explored and photographed, in and around, BLOWING CAVE which is located about 2/3 of a mile west of the post office in Cushman, Arkansas. There is a sharp right curve and, then, another sharp curve to the left, and on your right is a gravel road that runs west and about ten degrees to the north. It is about one & 2/3 mi. to the end of that gravel road. At the end of the road, you will see a small stream of water on your left, a very large log and a board gate with a posted sign. The water comes from the underground stream of the cave. I mapped the entire area in Blowing Cave which is a minor or sub-portal. As far as I can figure, we explored about 1/2 mile inside the cave. There are other small and large portal entrances to the west and south of this cave, but this one will lead you to an underground lake. When you find the lake, search on your right, and don't overlook anything around that lake, but do not come alone or unprepared. Somewhere near the lake are tunnels, and it is possible to locate man-made steps that will lead you into the corridors. Once in the corridors, follow in the direction of a light. It appears to be blocked of any light to the left.

In order to find the opening to the cave, itself, look for a small trail which you can drive up for about a short block, just before you get to the end of the gravel road at the stream. Then, follow the path which will curve to the south after you drive up the small trail. You will not see the cave until you almost run right into it. It appears dark, lots of shadows, and gives the appearance of a story from Dracula with all its mist, shrubs, trees, vines and hidden shadows.

When you reach the lake, it is very still, wide and deep, and the "four" used an old 3-29 tire inner-tube to get across it. So far as I can figure, it may be as far as 3 miles to the lake. The tunnel I was in runs to the east and, then, turns to the southwest. Be extreme cautious, for it is wet and slippery, especially when it is raining. Remember, two were lost in Blowing Cave and have never been found. So, have the proper attitude and you, too, can enter into the subworld and meet the Teros.

We took over 340 pictures, in and around Blowing Cave. I was drawn to a certain place where I examined a small rock, about four feet high, three feet wide and about one foot thick. It was very BLACK with some circles on it that appeared to be symbols. I took ten pictures of this rock. After they were developed, I found that all ten of the continuous frames were "overexposed," which is not possible under normal conditions. That rock had some form of unusual properties, and some crystal appears in its structure. Furthermore, while I was examining that rock, I did get a surge of physical energy which lasted for almost three weeks. Another thing, I broke out with some kind of rash which lasted for another two weeks. Evidently, the energy from that rock may cause some form of physical cleansing which causes the body to have a rash. Also, in another camera, there were seven continuous frames that were "overexposed." At the mouth, and in, the cave, I took seven pictures which showed a beacon of light that engulfed the entire area of the cave entrance. It does not seem to be a "one time" situation, for we had taken pictures of the beacon light from several different angles and, also, near the log, on each of the five different days we explored. Yes, there is an ancient artifact there, near Blowing Cave.

Remember, because I am able to take paranormal photography, I've proved that Blowing Cave is a "Portal" to the subworld. I brought back proof that the story of George Wight, and the four, is true, as well as bringing proof of our own personal experiences. One of my pictures, taken near the stream's exit tunnel, shows the face of a person wearing a helmet. Did the Teros, from the subworld, superimpose an image of themselves, to prove to me that they do exist and are sort of guiding me into the subworld?

Even though a person uses psychic abilities to probe into the subworld, they will never get any real proof. Nor will you get the true picture of the subworld, until all of us physically walk down the tunnels of the cavern and into the corridor. That is the only way we are ever going to know for sure. Eliminate the occult, the mystic, the crystal balls and the pendulums, for we do not know by that process. The only way to know is to get in there and dig, search and probe, now. The subworld is not occult or controlled by earthbound spirits. The subworld is solid, physical, material and populated by real flesh and blood people. They are different in some respects, but are real to honest people and not too much different from you and me.

Many fear the subworld, but it is not as it has appeared. Fear was caused by confusion and by the Deros (and earthbound spirits) that cause poltergeist activity of noise and confusion. In some instances, Dero will attempt to control an individual's brain wavelength with such as, "God told me to kill, etc." The Teros have controlled the Deros, ever since 1955, after a small war that destroyed many Deros and chained others that came from the southern cavern world. Most Deros are from the upper caverns, anywhere from the surface to as far as five miles down. And, Bigfoot, Apeman, etc., are really Deros. The white rays (Teros) are in complete control in Missouri and the Ozarks of Arkansas, Tennessee, North Carolina, Kentucky, West Virginia and Virginia. Forget the fears and confusions of the Palmer and Shaver stories about the cavern world. There are no man-eating Deros in the states where the white rays exist and are in control.

The "mystery" is over for now, because, for the first time, you have the information on how to get into the INNER HOLLOW EARTH. You do not need to go to the North or South Pole entrances. It is right here, in Blowing Cave, near Cushman, Arkansas. So, now, dig in. Let's go, obtain and learn the knowledge of the Ancient Immortals. The Ancient Immortal Gods are gone, but the Teros believe that they will come back by the turn of this century and take them away from this world.

The pathway of the Ancient Immortals is by way of the corridors to the hollow inner earth, and all of their artifacts, their wonder mechs and immortal buildings are all still there waiting for their return. The coming of the Ancient Immortal Gods and our entire heritage is at hand. Let's take it now, and go on to the stars and beyond this world of self-destruction. The Southwest Expeditionary Unit now has fifteen active members ready to explore with me. From what I can see, at this time, we will make our next exploration to this cave sometime next spring. So, who else is willing to join?

Map of Blowing Cave near Cushman, Arkansas.

CHAPTER SEVEN
Notes On Caverns
By Charles A. Marcoux
VOL. 3 NO. 3 – May, 1982

There is a story of a French farmer who had a cavern located on his property. Many times he entered the cavern and disappeared for intervals of more than six months. Each time he reappeared, it was found that he was aware of the latest news from Europe. In those days there was no means of communication which could relay news from Europe in less than six months. Later, it was always found that the news related by the Frenchman were verified by reports that later arrived from Europe.

In 1950, a couple of boys on bikes wondered by the mouth of this cave. The mouth of the cave was sometimes flooded with water from the spring thaws and rain. The boys found an old rotted boat by the cave's entrance, and after getting in, proceeded into the cave. Accidentally they found a narrow recess in the cave wall and after going thru it, discovered steps going down to the depths. Having no means to light their way, the boys returned to the outside world for supplies and equipment with which to continue their investigation of the steps. After gathering equipment and supplies, the boys returned to the cave only to find that the spot which they had hoped to explore had disappeared and could not be found. After relating their experiences to other individuals, the boys and several others attempted to again locate the narrow passage, but to no avail. Others of the Shaver Mystery Club also attempted to locate the hidden steps, but they too met with no results. The passage to this day remains lost and apparently is blocked by a ray. The cave is located in Kentucky, near the Virginia state line. This cave may extend into Maine and up into

Canada. It is possible also that it leads directly to the Northern Hemisphere and the entrance to the true North Pole. Legends claim that man came from the "North" and the Eskimos as well as Indian tribes of Canada also claim to be from such a land.

About the article in Vol. 3 - # 1 in *The Hollow Hassle*, entitled "The Unknown In Wyoming," there is a Medicine Wheel in the Superstition Mts., near Apache Junction. It appears that the Indians know nothing about it. These Medicine Wheels are astronomical observations and were made by the Lemurians, after it sank below the ocean. The people of MU say the coming floods, and many colonies were sent to America. They were not "Indians" but a fair, white race of people about 8 to 10 feet tall, who were not "WARRIORS", but peaceful. The Indians came later and are of Mongolian descent that came from Asia before the flood. The Indians came to Central America and invaded the Maya culture and the Mayas (white, blue-eyed people from MU), being a peaceful race, had to flee to the North. Some still exist in the Cavern World near the four corners, while many others went further north where other Indians came later to destroy them. That is why the Indians don't mention them, but some Indians are from MU, sort of an off-shoot from mingled races. They were friendly until they found they couldn't cope with the coming warriors from South America. So, they too became warriors, and in so doing, made war against the Mayas. They are guilty of their own crimes against the Mayas, and that is why they won't mention, or why they deliberately forgot about the Mayas.

I personally believe that the Mayas, in Wyoming (and the Medicine Wheel), were pretty well established in deep caverns with a well-fortified city (citadel), or were well in the progress of doing so when the flood came. The Navajoes tell of the white, blond giants along, and near, Route 666 (AZ. and N.M.) and near Tipon, AZ. A friend of mine, Lee Summers, an Indian himself, often spoke with the Navajoes and Zuni who told about the white, blond giants. Lee Summers (deceased not) was a Deputy Sheriff in Holbrook, AZ., and knew the Navajos well, and the Indians don't bother the giants, even though they often encounter them while near this mountain.

I believe that the Mayas came to the Superstition Mts. and retreated into the ancient cavern of the Lemurians. It is entirely possible that their descendents still live in the deep cavern under the Superstition Mts, but are now of a mixed blood of many races. When the Mayas fled the invaders about 13'000 years ago, they did retreat into the deep caverns and adopted the ancient cities of the Lemurians and the Atlanteans (known as Azaltans or Saturians) and are in a secluded area protected by the high mountains about 85 miles southeast of Mexico City. The Margaret Roger's story (Amazing Magazine 1947-48) tells where she was taken to the high mountains and into a city in a cavern far below the earth. These people were friendly and peaceful, with a super-science that can only be achieved by thousands of years by the Ancients. Did Margaret Rogers encounter the Mayas (Mayax) and possibly the Naacals, who were the priests from the seven temples (Cibola) of MU, and did bring forth the Maya empire? (The buffalo skulls show symbols of the "HORNS" of the Buffalo heads, which mean the horn, a crescent or quarter moon. The priest sits in the center between the two points of the horn to give counsel, messages or a direct voice from the Great Spirit, etc.

About the Tongue River Cave near Sheridan, WY., as mentioned by the Hefferlin Manuscript: The ancients had "RAY MECHS" which were able to pack gravity into rock. That is, they would harden the rock and make the tunnels harder than iron, or steel, by the use of packing gravity into the pores of the rocks. Many of the old tunnels leading into the caves or caverns were made secure by this means, and the walls were glassy smooth to reflect sound or carry thoughts that were also reflected back by the glassy surface of the tunnel bores.

The reason so many tunnel bores of the Ancients were "sheared off" was due to the sir-king of Lamuria and Atlantis (Azaltan) and the raising up of the lands of America and other continents that now exist. Most of the ancient cities that were underground were destroyed, or nearly completely destroyed, by the flood and the sinking of these two nations and the final tilting of the earth. That is why most of the Lemurians, Titans, Norsemen and Azaltans were almost completely annihilated. The races' source of supplies were suddenly stopped and cut off, and they eventually became warriors, murderers and cannibalistic. We have not yet recovered from that flood and the sinking of the two nations. Those races that survived became warriors, and we follow in the footsteps of those survivors by being warriors, destroyers of life and property, and we are being led into the same situation of facing our annihilation and, maybe, the end of the human race. I am not presenting a fear thought of world destruction, for it is quite apparent that certain world politicians are leading mankind to his demise and the end of civilization as we know it. At any rate, my opinion is that because of the attitude of the human, he deserves what he gets, and I can care less if I survive on this earth. The real object is that it all points to the fact that the "Source of all Life (God, or whatever it may be) wants to secure mankind so that his seed can survive, as it says in the Book of Revelation and in other writings.

There are lots of entrances to the inner earth in Colorado, and I am sure that under Denver is a direct opening to the underworld, which is close to the entrance on Wilford South's map. (See H.H. Vol. 2-#1). It is covered up by officials, as usual. Also, I feel there are entrances at the four corners, Silver City, and some further northeast in Colorado. The Aztecs had mining interests in Colorado, as well as citadels (forts) in deep caves with the Teros and Deros in the underworld.

TILL NEXT TIME...

Blowing Cave, Cushman, AR

Charles Marcous

Lorene & Charles Marcous, Mary J. Martin

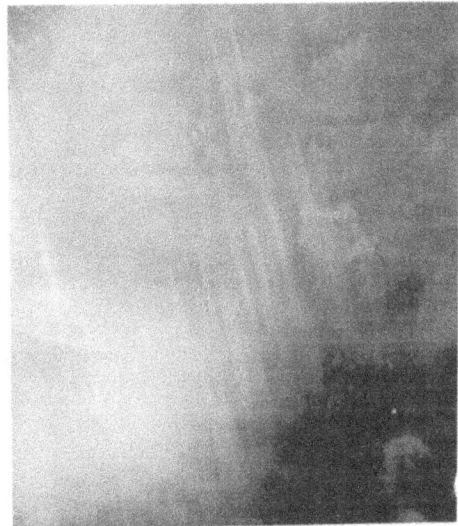

Mysterious rays at mouth
of Blowing Cave

RICHARD S. SHAVER

DOT - DICK & NEIGHBOR JEFF
5-21-75

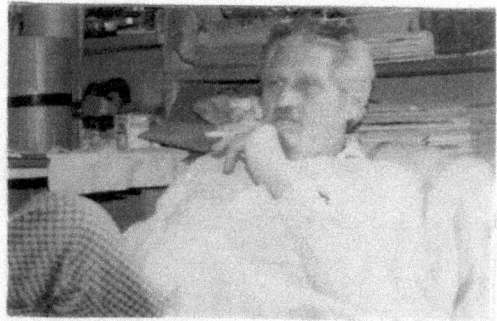

5-21-75

BILL Crout   DICK

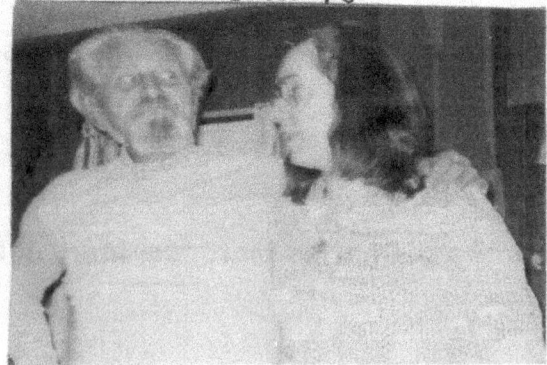

DICK   MARY Martin

MARY Martin   DICK 7/75?

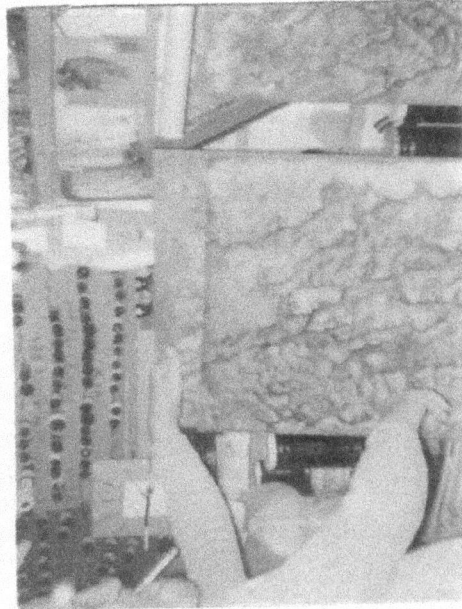

LABRON - TOM - DICK - MARY
Bynum   7/75?

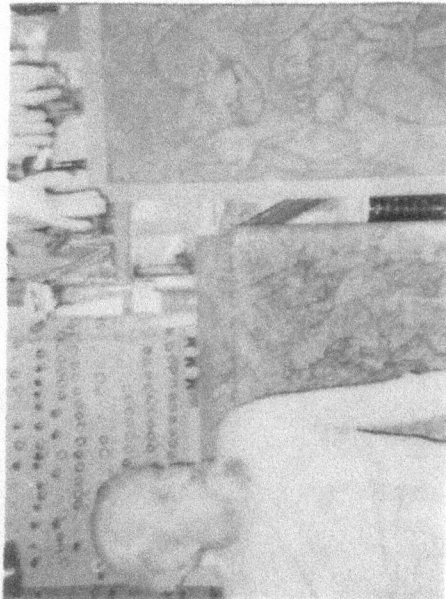

Dick & Rock Book Paintings

CHAPTER EIGHT
Charles A. Marcoux's Search For The Inner World
Lorene Marcoux
VOL. 5 NO. 3 – May, 1984

Is there really an inner world as Richard Shaver claimed in his stories that came out in Amazing magazines many years ago? Charles A. Marcoux was one of the first to say "YES." Then, he proceeded to spend the rest of his life making every effort to try and prove it. In his own mind, he had no doubt that there was a cavern world beneath the surface of our earthy because he had been there to see it for himself.

When Charles was a very young child, a tall man would appear in his bedroom late at night and take him by the hand to lead him into the cavern world. They would float along until they reached a large building that had many wide steps and pillars. This tall, kindly being would take him inside where he would be handed a large book. Charles would look at this book and then, suddenly, find himself back in his own bed, unable to remember what was in the book. Charles was visited by this same tall being many times and taken to the fascinating world of the caverns where he was learning many things from the books they kept there. Even though Charles was unable to remember anything he learned upon returning to his bed, he seemed to sense that the knowledge would be there in his mind whenever he might need it.

Upon reading Shaver's stories about the cavern world, Charles was very excited about the information contained in them because he had finally found confirmation to the puzzle which had been in his own mind since his early astral trips to the caverns. At last, he had found the missing piece of the puzzle!

Some years later, Charles moved to Phoenix, Az. where he spent years searching for an entrance to the caverns in the Superstition Mts. During these years, he talked to different people who had lived in the vicinity for a long time, and they told him many legends and unusual stories which helped to convince him that there was definitely an opening to the cavern world located in this area. While hiking in the mountains in search of an entrance, he often saw UFOs in the sky, day or night. Besides this, he had many interesting experiences in these mountains which he wrote about and had published in a number of publications.

But, he never found an entrance to the caverns even though he felt he was very close to doing so a number of times.

Actually, Charles managed to get better proof of the existence of UFOs, instead of caverns, through endless photographs which show strange phenomena, lights, force fields, UFO, spirits, etc. As a matter of fact, his unusual ability to have phenomena appear on a high percentage of every roll of film came to the attention of a local TV station in Phoenix. They presented two special features about him and his photos in Sept. of 1983 on KTVK news programs. His photo evidence showed that there are other dimensions existing, side by side, with our own physical dimension. These photos presented proof of what he had known to be true all along, that there are other worlds existing on, in and above the earth. What he had been seeing and hearing through his own special psychic abilities were showing up more and more frequently on his rolls of film.

Finding proof of the existence of UFO on his photographs only encouraged him to feel that someday he could present evidence of the cavern world, also, eventually. As he often said to me, "UFO, spirits, ghosts, cavern worlds, telepathy, healing and all the other mysteries are all connected, as they are only different aspects of one truth. Man may try to separate these aspects into different categories by limiting himself to studying one particular subject, but when truth is discovered about that subject, it will present another question to be searched into."

As you know, about 50% of our nation has come to believe that UFOs are real, now - this is a very substantial gain over a few decades ago. Almost half of the people are now convinced that there are UFOs. Still, the idea of a cavern world seems to be nearly impossible for the majority of the masses to accept, yet. However, just as more and more people have come to believe in UFOs, they will also accept the idea of a cavern world, someday. Of course, just as there has been a cover-up about UFOs over the years, there is a cover-up about the cavern world, too. Those who wish to control the minds and lives of the people would have much to lose if the masses discover that there is a place below the surface of the earth where there is a better way of life. This is especially true about the men in the black robes who are the priest class, the law makers and the money changers (usury). After all, the Tero subworld doesn't have any of these black robes in their world, and they get along just fine without them. Neither do they have many doctors as we do, for they have the means to heal various conditions very quickly. Because they have very little crime, they also have no need of police, jails, court systems, etc., as we do. Therefore, those in "power" on the surface do not wish to have the people learn of new ways that will result in taking away their own power. So, there will be a continuing struggle to cover-up the existence of an inner world by those who have the most to lose.

In 1981, Charles received a manuscript which told of George D. Wight's trip to the cavern world and how he lived with the Teros. Charles felt he would like to find out for himself what life is like in the subsurface world by going there, himself, in physical form. So, he presented this story of Wight's experiences, by writing about it in his column called Straight Talk in the *Hollow Hassle*, while preparing for a trip to the inner world as soon as he could manage it. He hoped to bring back some sort of proof to give to the public to help convince them of the reality of the inner world, just as Shaver had brought back the Mantong alphabet as proof of the cavern world. The Mantong alphabet was carefully researched by Charles and others at that time, and they were convinced that it was worthy proof of Shaver's claim that he had gotten it from the cavern world. The odds of Shaver being able to concoct the alphabet out of his own head and have it turn out to be workable with our own language is beyond belief. It had to come from a higher source than surface humans. Therefore, Charles, too, planned to bring back some kind of proof, if he possibly could.

Charles' effort to explore this cavern world for himself was unexpectedly ended in Sept. of 1983. He suffered a sudden heart attack and died while out searching for the entrance to the cave that Wight had used to enter the cavern world. True to the sign of Aries, Charles was a pioneer who explored many 'mysteries" by searching physically and by listening to the "voice" from within that guided him along to reach his objective. Now, it is being left to others to take the next step in the ultimate plan for the future of mankind. Charles Marcoux did his small part by presenting the truth as he knew it about his search for the inner world and left his footprints in the sands of time upon this Earth.

**Charles Marcoux Being Interviewed by TV News Crew
July 3, 1983.**

CHAPTER NINE
Grain of Gold
By P. Martinez
VOL. 5 NO. 3 – May, 1984

"All these stories do not agree together...Yet in every one of them lay hid something of the truth, a grain of gold in the ore of fable that might be found by him who had the skill and strength to sook." -- *Ayesha, from H. Rider Haggard's WISDOM'S DAUGHTER.*

Charles Marcoux was on the brink of a great discovery when cut off by untimely death in September of 1983. There at the portals of the unknown, he was stopped dead in his tracks, literally. This study of the ancient breastplate is a tribute to his memory... In the October 31, 1982 issue of the Hollow Hassle Hotline, the late Charles A. Marcoux continued his chronicle of underground exploration. Going into some details of a subterranean encounter that occurred some eleven years ago, Marcoux describes a chest device worn by our underground neighbors, the Teros. The chest piece and accompanying wrist device enable the surface visitors to communicate freely with the Teros, who also transfer thot to thot. The Teros, known by other names, are generally considered an ancient race, dwelling in the unknown hollows of earth... In the sacred records of the ancient Levant, as well as the Americas, there are breastplates, designed for wear on the chest. In the old world of Egypt, Palestine, Tyre, etc. Samples of such relics are known especially the breastplate worn by the high priest. The best known of these was called - the Breastplate of Righteousness and Prophecy worn by Aaron, High Priest of the children of Israel.

THE BEST OF THE HOLLOW HASSLE

There is a strange and arcane figure wearing a breastplate (see illustration) like a tablet which was last seen in the Matto Grosso of Brazil in 1925. The relic is a 10" black basalt figurine which gives off a "peculiar electric current." Inscriptions of some sort appear on the chest piece and also across the ankles.

The breastplate, or Ephod, worn by Aaron also contained engravings which rabbinical study identifies as the names of the twelve tribes of Israel. According to this tradition, the stones also lit up in miraculous response to questions, an oracle.

But, laments the Jewish historian Josephus (1st Century), two hundred years earlier, the stones ceased to glow, having gone out in the bitter winter of the children of Israel. But this was many years after Aaron's time, perhaps a thousand. Aaron, we know, was the brother of Moses, otherwise known as the Egyptian Seti-Meshu. Moses acquired his mastery in the land of the pyramids before leading his people and their sacred relics out. An Egyptian prototype of the Ephod was known as Re and Theni, meaning Light and Truth, the truth of righteousness and the light of prophecy. The Egyptian breastplate was worn, in this case, by Themi, the goddess of Truth, whose better known cousin is the Greek Themis. The Caliph of Babylon discovered such a breastplate while poking around the pyramids. On the breastplate, which was worn by a mummy, were jewels set in gold. The man's body was found inside a statue which in turn was placed inside a hollow stone. On his forehead was a carbuncle, the bigness of an egg, shining like the light of day. This was at Cheops, which, considering the legend of their transference to Central America is curiously cognate with Chiappas.

Talking of the Americas brings us closer to Marcoux's Teros and their origin. The term Tero has an ancient cognate in Therapeutae, which has resulted in the words therapeutic, therapy, etc. The Therapeutae were healing men, ascetics who dwelt in ancient cave temples and retreats. They are known from Essene times and are named by the same root inherent in the words holy, heal, whole. Terah, incidentally, was also the name of the father of Abraham, progenitor of the Jews. There is yet another ancient tera- which applies here.

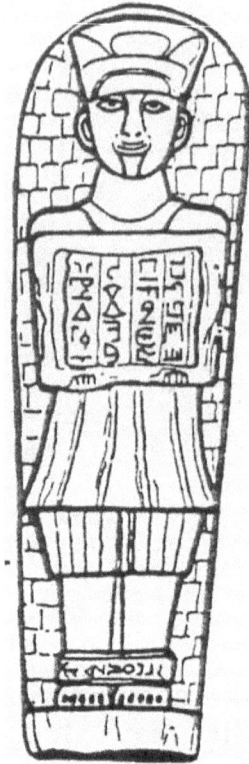

The Bible mentions teraphim, images which answered questions (Ezekial, Genesis). As the light of the stars filled the carven statue, so states Maimonides, it was put en rapport with distant intelligences. In such manner, the teraphim were able to teach "arts and sciences" to men of the earth. According to Seldenus, golden texaphim were known also to the Egyptians. Still other teraphim were in the form of heads, able at times to speak magic words which had been engraved on them.

The underlying meaning of tero/tera-suggests a source, known to the ancients, of truthful, if not prophetic, answers. The term further suggests a close link between healing and the arts of communication. Marcoux, in describing the Tero's chest device, calls it telepathic. Along with this unusual sensation is a feeling of "peace and comfort." Further down in the realm of the Teros, some 15 miles into the earth, there was a "Reeling of physical energy which gave some form of stimulation to our mind."

Returning to the Hebrew prototype, Aaron himself taught that the breastplate was an instrument for learning the will of God. The complete garment, known as Ephod, was much like a vest, fastened at each shoulder with a large engraved onyx button. As an oracle, it was known to answer "yes" if the button on the right were illuminated; if the left, it was "no." The simple Yes/No oracle is familiar, not only from ancient records but also from spirit phenomena of modern times. The famous Amherst mystery, one of the best documented poltergeist cases of all time, includes several incidents of spirit rapping, in which two-way communication preceded with ease as long as a Yes/No format was followed. A simple code was used, one rap taken to mean "no," two raps as "doubtful," and three raps a "yes." The spirit answered correctly such simple questions as how many people were in the room at the time.

The Amherst case generated some interesting theories, such as that of Dr. Edwin Clay. A Baptist clergyman close to the case, Clay concluded that the weird phenomena were electrical in nature, the afflicted woman having become a "human electric battery."

The doings of mischievous poltergeists and the will of God may seem strange bedfellows. Indeed they are strange, but the history of early priestcraft is no stranger. One hears much talk of the difference between the black arts and beneficial magic. Yet, upon closer inspection, the distinction falls apart. "The Craft" for instance is a term used both by the Masons and Black Magicians. The forces of light vs. dark meet again in the history of such relics as the famed crystal skull. Like other oracular heads, this relic too could be set moving and talking. The high priests wielding this magnificent artifact made use of a prism in the cranium to magnify light, setting the eyes aglow. Pipes projecting upward from the base reflect light from the eye sockets; the moveable jaw easily coordinates motion with speech. The most obvious function of such devices, neglected by most theories, is their ability to create a spectacle. It is precisely this ability that made them so powerful and so secret.

The early priests of Greece had such a relic, fashioned by Hephaestus, the Blacksmith of Olympus. These were two golden statues which could move on their own accord. Likewise in the temples of Thebes (Egypt) there were images of gods which made gestures, could speak, and were most likely operated by priests from behind the scenes.

This is not to suggest, however, that trickery was the basis of ancient oracles simply that secrets were kept by the priesthood. More important than the question of fakery is the question of motive. Mme. Blavatsky tried to show this a century ago, but she was misunderstood. In any case, the usefulness of awe should not be underestimated. Nor should motive. Josephus more or less blames the militarism of the Hebrews for the loss of the magic in the gems. For the oracle had ceased to evoke the will of God (Jehovah). In its heyday, though, God would declare victory in battle by lighting up the stones of the breastplate; even the enemy could see the glow - awe and intimidation at its most potent. Another relic of the Hebrews, the staff of Moses, was also used to assure success in battle.

These items were not mere "ritual" paraphernalia, as the historicity of the Bible will prove; not too different than the story of Schleimann's Troy or Edison's gramophone. The history of disbelief is too far afield for our subject, but a healthy look at its pages is sobering. Modern occultism has turned scientific and now wishes to explain these awesome forces in natural, scientific terms. The meeting of the minds has not yet occurred, yet the invisible world comes ever close into view. From this perspective, the only myth is that the Scriptures are a myth

Moses, for example, was instructed to remove his shoes before ascending thru sacred ground of Mount Sinai. Mere ritual? Bible historicists and super-scientists would have to agree, there was a good reason for God's request, more than mere formality, and more than symbolism. Our god is not supposed to be a jealous, petty, or vain creature, given to supercilious protocol. He is a good god and can reach man through his inner resources. Is this not exactly what he did?

THE BEST OF THE HOLLOW HASSLE

Moses did remove his shoes and in a burning bush that did not burn, received the Law. How did enlightenment come? Wouldn't we like to know! Still, we do know that an overwhelming energy field came down the mount with Moses, surrounding the prophet with a glow that only gradually faded away. We must not be satisfied with symbolic interpretation of such events, for just so we miss the point.

The sacred relics of the Hebrews are succinctly mentioned in a legend from the hexameron. On the second day of Creation, there were created the well by which Jacob met Rebecca, the manna which fed the Israelites, the wonder working rod of Moses, the ass which spake to Balaam, and the shamir. Rabbinical lore further teaches that Moses, and later King Solomon, had recourse to some of these items, particularly the miraculous shamir, which cut stone "where iron does not bite." When Moses wished to engrave on the stones of the breastplate the names of the twelve tribes of Israel he had recourse to the miraculous shamir. According to one version (Ginzberg, *Legends of the Jews*) the names were first traced in ink and the shamir was then passed over them; now the traced part became graven. "As proof it was magical, no particles of the gems were removed in the process." What magic cuts stone without waste? E. von Daniken, although much debunked as a Sunday archeologist and worse, offers a simple theory; the thermal drill, he says, behaves in such a manner. Investigating possible methods of tunnel boring, the Swiss writer boldly proposes:

"...they were presumably equipped with a thermal drill, developed at U.S. Laboratory for Atomic Research at Los Alamos. The tip of the drill is made of wolfram and heated by a graphite heating element. There is no longer any waste material from the hole being drilled. The thermal drill melts the rock and presses it against the walls, where it cools down."

This author also believes that only electron rays would explain the uncanny "glaze" left in those vast tunnel networks of Ecuador and China. To follow this argument is to end up with shamir equaling, for all intents and purposes, the rod of Moses, which is also interpreted as a kind of ray gun, most like our laser. The second day of Creation was certainly productive!

Notwithstanding such worldly applications, these creations were originally willed by God for the advancement of man, and that means ALL men. Here again is the deciding factor - motive, good judgment. What a mess we have made of His creations, making Him to witness so many misapplications in all his great kingdom! The black and beneficial arts are just about inseparable, as an instrument of knowing can easily become an instrument of power.

But let us return to these graven images. Supposing the engravings on the ancient stones were of critical importance, graven circuits designed for an electrical response. Scholars can't quite agree on the meaning of the twelve stone of the Essen, or Hebrew breastplate. One theory even proposes that the twelve symbolize the Ten Commandments plus the Lost Two. In due course, the last two will be revealed, pending man's upward progress through the grades of wisdom and law.

But suppose these, and other mystical engravings were electronic circuits which created a current or wave with remarkable effects. Computers have recently helped translate some of the ancient seals. The Star of David, for example when used as a printed electrical circuit, creates a standing wave, which, incidentally is also produced by the rod of Moses. All of this, too, is highly descriptive of the spiral movement of energy through the "mysterious" monuments left on the earth in bygone ages. Do we look for a literal radio set, or do we interpret sacred records as given: Man was instructed to stand upon the bare earth and receive the truth unto his readied spirit. From the Book of Aph, Son of Jehovah, (OAHSPE) the inner value of such circuits or magic words is given:

"Now, I, the Lord, reveal in this, the kosmon era: My angels abode with the chosen in the days of the aforesaid sacred scriptures. And when the words were repeated for the stopping of blood, behold, my angels compressed the veins. Not the words (sic) stopped the blood, but by the words mortals became in concert, with my hosts...without such words there could be no concert of action betwixt mortals and angels. Think not that the mumblings of words by ray prophets were without wisdom..."

Moses was obliged to contact earth with his bare feet for the proper circuit of energy to pass through his body. This differs only in detail from assorted out-of-body practices of the gurus and pundits of India. The ancient practice of anointing with oils also encouraged the free passage of subtle energies. In the Amherst case, noted above, Esther Cox became victimized by this uncontrolled energy. Because it was believed that her trouble was due to electrical discharges, her closest associates insisted that she insulate herself from contact with the ground by wearing thin glass soles inside her shoes! (She discarded them after they made her head ache and her nose bleed!) But her manifestations were frightening; among the many observed, "I have often watched her to find out how she came downstairs, she seeming to fly."

King Solomon, we know, was wise, but not wise enough to build his temple without the sound of iron. His wise men then drew his attention to the stones of the high priest's breastplate, "which had been cut and polished by something harder than themselves." That something was shamir, but to obtain it, Solomon had to summon Asmodeus, king of the devils.

We find, now, in King Solomon's book of evil spirits, that to summon Asmodous, the exorcist must "stand on his feet all the time of action." Furthermore, the seal of Asmodeus "is this, which thou must wear as a Lamen upon thy breast." Solomon, Moses, Aaron were all "charged" with the same task. Each was a master, and by means of his mastery, grounded his being to the full circuit of heaven and earth. In olden times, when the gods were numerous, mastery meant rapport with the angel's prophecy, a fully illumined heart. Today we have lost track of mastery and angels. But who, other than man and man alone, can translate the will of God? How adept one is for this awesome task is another matter. Moses didn't think he was up for it; but he was made to serve, and handsomely assisted.

Joseph Smith, founder of the Church of Mormon, was a modern Moses. While still young he was visited by the Angel Moroni and given to locate and translate the Book of Mormon. Like Moses, he was assisted in the difficult work. For along with the golden plates, he discovered a lens-like object, a crystalline device. To possess these stones was to become a seer; their purpose - to translate the Book.

Occultism means well in its current efforts to de-mystify the sacred relics. It establishes, for one thing, the fact that mysteries often have reasonable causes. For another, it is the natural counterweight to the stalwart position that every tiling is symbolic or ritual in purpose. The sacred manna, for example, is believed by the technical occultist to represent the honey-like secretions of the pineal/pituitary gland which are activated by the standing wave of energy. To the orthodox scholar, however, all that can be admitted of these items - the manna, the tablets, the rod, etc. - is that they "spoke of newness of life with God." Curiously neither is wrong, for in the god-realized state, we have the fulfillment of all hunger. Curious that one of the many interpretations of the rod (vril) is given as literally meaning "food for all." Let us not debase the wisdom of the ancients with an all-too-eager science or all-evasive symbolism. Neither technology nor literary eloquence will ever resolve the problems of man.

Neither should we overestimate the wisdom of old. All humans need help. The gold of the gods, apart from its great splendor, is, of course, a semi-conductor and served our forbears in the transmission of divine "truth." The golden sandals of Montezuma; the oracles of Hephaestus and the Egyptian teraphim, made of gold; the golden sockets containing the gems of the high priest's breastplate in Israel and Cheops;   the golden tablets given to Joseph Smith by the angel Moroni;   and on and on.

One problem remains,   and that is the nearness of the White Lily to the Black Orchid. The Breastplate of Righteousness & Prophecy   is found in the Americas, but lo!, it   portends blackness and corruption.   In place of Light,   the New World replica is considered the remnant of an empire which once held absolute mastery of the world, with alarming use of the black arts, "long prior to the rise of Egypt."

One of our greatest Americanists, in a fantasy exploring this unknown civilization, depicts the repulsive image of their sun god, a bestial human thing with hideous eyes, gleaming fangs, slavering tongue.   In his clawlike hand he holds a human heart. The image is carved from a single great block of black stone, and covering the chest is an immense gleaming gem-studded disc of gold.

All this is to show that there is no lily white history of man, but only a long and slow .retreat from mindlessness and savagery. Our age-old breastplate has shown this, but there is yet one chapter to be written, and that is initiation into the depths of the Earth. The Tero breastplate, with its telepathic and energizing functions, has its analogue not only in the priestcraft of early times, but in inner Earth lore as well.   In the remotest of ages, we have a prototype in the black figurine (illustrated), alleged to belong to a lost subterranean city of Brazil. The vast tunnelworks connecting these underground forts may have been built with the aid of an erstwhile laser technology. The   same   complex   of awesome forces will,   in the fullness of time,   help to demystify many of the sacred relics - the breastplate, the rod of Moses, shamir, and so on.

The   underground   connection   is   next   impressed   on   us   in biblical times where the relics are "buried," only to reappear two thousand years later in North America (Mo monism).   Here enters the question of the Lost Tribes of Israel. For over one hundred years researchers have been investigating the possibility that the tribes disappeared into the earth,   especially into the unknown hollow beyond the ice barrier of the North Pole.

An underground cult of Greek times that of Hephaestus, or Vulcan, again directs us toward the subterranean unknown.   The oracular images/statues credited to this deity forge a strong link between the origin of the breastplate and the history of metallurgy, volcanic action, and the mysteries of Mt. Etna. Moving into early Christian times the cavern-dwelling Therapeutae suggests another prototype of the sub-world Teros. Who knows how extensive their cavern community? Latter day archeology around the Dead Sea may have some answers, just as its discoveries changed history only a few decades ago. But seeking the favor of his fellows, there will be no archeologist with enough grit, no intellectual smart enough, no historian learned enough, no specialist with sufficient expertise - to lay bare these mysteries. They will always be just beyond reach. For we are retracing the steps of magical times. Enchanted times when a core of humanity, the heaven-born of men, wielded awesome powers through both personal mastery and knowledge of occult science.

Today we honor neither mastery nor any science other than our own. But still we fear the spirits of darkness. Our hold on goodness and virtue is tenuous, negotiable. The breastplate of righteousness and prophecy is gone; it seems, except perhaps in some hidden sanctuaries of Earth where the past is preserved for us by unknown guardians.

Is it too much now for our pampered egos to surrender to the possibility of subterranean dwellers of earth? Why must a perfectly good planet be inhabitable only on the surface? What arrogance allows us to close the books on a realm that has never been explored? What holds together our argument of solid Earth? Deductions, guesswork, intimidation, ridicule, habit.

Still, in the chartered and unchartered fields of learning, the true elite, the upper grades in all human studies, the mental giants, known and unknown, the masters of their craft, those who do love the truth, these, today and yesterday, wear the breastplate of righteousness and truth, in their own right. Those who, unfailingly, turn our attention to the marvels that govern life on earth, those who incite a healthy respect, and curiosity, for the unknown, those who challenge us to know more.

We have taken one detail of inner earth lore, a magical chest device which, in its many forms, predisposes the wearer to a finer plane of communication/thot. That plane partakes of the electric/(magnetic) kingdom, to which we humans belong. As with any myth, the historicity of hollow Earth must yet unfold. I have been told by one who has studied hollow Earth for forty years, that 70% of it is rubbish. A wonderful statistic! Leaving us sleuths with a fascinating 30%. So, to work! And let us neither overestimate, in idolatry, nor underestimate, in skepticism, the miracle of life, for to do so would be to betray the true and natural marvels of Self.

THE HOLLOW HASSLE

CHARLES A. MARCOUX

VOL. 3 - NO. 2   FEB. 1. 1982

CHAPTER TEN

"Astral" Inner Earth Kingdoms Are Engaged in Conflict Over Us
By Tal LeVesque
VOL. 3 NO. 2 – February, 1982

There are many levels of being that interpenetrate and influence Earth existence. The Earth plane provides a predominantly three dimensional existence, for it is a world of matter and form, but the dimensions overlap.

There are benign factors inhabiting these various spheres as well as many that are decidedly malignant,

The False Lord cast the astral Wo-Man race with its awakened awareness, out of the Garden, onto the planetary surface, to live in flesh bodies of the third dimension. From flesh of astral dust they were cast into skins, and became born into death, sorrow, pain and duality. He did not create soul-mates, nor divide each soul, but rather created an unbalance of magnetic polarity in each being. This caused the male and female to seek each other for fulfillment and balancing of their magnetic beings and the Dark Lord used this process of attraction and repulsion to draw energy for his sustenance.

The Serpents of Wisdom and Higher Awareness are Etherial Teachers, neither of astral dust nor skins of animal bodies. But some Serpents dared to incarnate in mortal form to teach, even to this day.

It appears that the Inner Land, Middle Earth or the place called the Garden of Eden, is not entirely rid of the Astral Hordes that first set out to control the lives of men. But for the most part, these dark powers are residing in the cavern worlds, midway between the surface earth and its interior. Great tunnels are known to exist throughout the planet and caverns of great size house the hostile saucer ships and other similar vehicles. They have caused a great deal of confusion among the surface dwellers. People become confused in trying to discern which are real space visitors and which are planetary powers of the negative forces.

The Dark Lord farms the surface dwellers, as an ant colony farms aphids. He started upon a campaign of cruelty against the surface-bound Wo-Man race. They were programmed and chemicalized to take orders, to kill and burn flesh, and do violence to all nature. For in these acts there was a kind of energy created which the Dark Force fed upon. He instilled in them the craving for blood and flesh, especially ordained slaughter houses he called his temples. Even Hitler fed the Dark Power the energy essence of violence, by sacrificing Jews.

Beginning before the time of Solomon, the astral overseers of the planet would sometimes bring their vehicles into plain view. This served to awe and frighten the natives into conformity.

When galactic souls incarnate as teachers, sane of them became entrapped in the molasses vibration of dense Earthly matter. But there is that latent restlessness within them, something missing, something sorrowful.

All races of Earth have a portion of the two tendencies, GALACTIC and insect-like, and either one or the other trait becomes dominant in each individual. The latent seeds of Galactic Man lie dormant in ALL of us.

UNDERGROUND UFO BASES AND WAY STATIONS IN THE UNITED STATES

Alien beings have established bases in selective areas around the globe. In central Texas's Robertson County -UFOs have virtually become nightly callers. Along with related ground level phenomena, including: Unexplained tracks on the ground; Mysterious deaths and disappearances of livestock. The interference of normal radio reception; strange code being received on citizens band radio; power black outs; etc.

UFO's are using the area near Calvert, Texas as a base or way station. Caverns exist beneath farmland on the outskirts of town.

"There is a complex network of Caves and Tunnels which connect somewhere underground. A check of geographical survey maps will show that Calvert is built directly on top of a fault line which zigzags for miles in all directions."

The ranchers and farmers in this area have reported hearing peculiar noises coming from deep beneath their feet. "Individuals living five or six miles outside Calvert has repeatedly been driven out of their homes into the cool evening air by the sound of generators. It appears to them as if a steady droning noise is originating from all directions but is loudest when ears are placed to the ground."

This bit of information has led me to conclude that UFOs operating around here have established bases for themselves far beneath the Earth's crust.

Going a step further, UFOnauts are probably obtaining the necessary energy to power their ships arid bases by making use of existing underground bodies of water.

"This portion of the state is crisscrossed with hot mineral streams that flow silently deep inside the Earth. Placed under extreme pressure, these streams could make one hell of a hydrodynamic generator, supplying any space visitors with all the energy they need and without having to construct anything at all above ground which would give their presence away."

Southern California, as well, has UFO, Sasquatch incidents and strange underground occurrences. One of the most significant took place in April 1973 when three students camped in the San Gabriel Mountains on the western edge of the great Mojave Desert "felt uneasy" and suddenly saw a gigantic man-like figure that frightened the campers.

The next day, three-toed footprints measuring 18 inches long were found in the hard-packed soil of the area. The UFO Research Institute (UFORI) in Redondo Beach, Calif, was called in for scientific evaluation.

After extensive field work in the campground area, UFORI discovered the sound of geared machinery and what might be compared to hydroelectric plant noise coming from "beneath the forest floor." While beyond normal hearing range, the subterranean sounds were picked up by special microphones placed directly on the ground. Nearby the remote campground is an abandoned silver mine which cuts through the mountain. One end has been closed by landslides and the other end sealed off by forestry personnel to protect explorers from the dangers of cave-ins. Sounds of operating machinery have been recorded coming from within the mine AFTER MIDNIGHT. Sightings of UFOs in this area are common place.

We are dealing with a superior technology with all the hallmarks of a military operation on a very large scale. Though most of the UFOs sighted are from the Inner Earth, for the simple reason that this is their home base, there are other ufonauts keeping a watchful eye on us.

I refer to those that come from other planets and galaxies. Perhaps our planet will become a battleground for two opposing factions of UFOnauts!

OUR PARADISE INSIDE THE HOLLOW OF THE EARTH

"He stretcheth out the north over the empty place." - Job 26:7 "I will sit also upon the mount of the congregation, in the sides of the north." - Isaiah 14:13

Over 100 Bible verses teach that the world is Hollow. These verses speak of a place UNDER the the outer shell of the Earth, Every verse that speaks of UNDER the Earth refers to Paradise in the Hollow Earth. A CENTRAL SUN lights up the whole Interior of the Earth.

The whole interior of the Earth is called "Eden," the garden of God. The word "Eden" means "Paradise." It is a grand enclosure or garden, out of which our race came from.

Paradise Lost & Paradise Restored, we are on our way back to Paradise inside the earth. We shall fulfill the covenant; when we go back into the garden; into the abode of LIGHT, wherein there is NO DARKNESS (i.e. No Night). You see, the present surface world is not our home. We will inherit our true home land...inside the Earth. In this lovely Paradise there is a Sun, a glorious Sun that shines with PERPETUAL LIGHT. It is a stationary Sun that never moves, never rises, and never sets. In this glory land it is one continuous day, for there is no night there.

From history and tradition we learn that there are several SUBTERRANEAN PASSAGES or Great Wide Tunnels which lead from the exterior to the Interior of the Earth.

Pindar, the Greek poet who lived from 522 to 443 B.C. said: "Happy the man who descends beneath the Hollow Earth, having beheld these mysteries, he knows the end, he knows the Divine Origin of Life."

Plato writes, concerning an Interior Sun, Apollo of the Inner Earth: "He is the God who sits in the Center, on the Navel of Earth and he is the interpreter of Religion to All Mankind."

In ancient Pagan times the superior race who inhabit the earth's interior frequently contacted surface dwellers, and temples were built for their occupancy during their visits. But with the establishment of the Church of Rome, these temples were destroyed and the existence of the "gods of the Underworld" was denied. Christian priests taught their followers that they were devils, evil beings who should be avoided, and that they should worship only as the Church-State, who 'wanted to control them, taught.

Thus the existence of a Subterranean World and its inhabitants became a lost memory. And just as religionists taught that it was an inferno of everlasting fire, so scientists preserved the error in their theory that the earth has a fiery core, basing this on the flimsy evidence that the further down one goes, the hotter it becomes. However, this evidence has now been proved false and scientists are looking more closely at the Hollow Earth theory, as they find that most things in nature are Hollow.

Every prophet, however great, must be initiated. Christ and his mother were members of the Essene (is-sonir - sons of ice) community. As an Essene priest, he could not have instituted the neo-Christian sacraments attributed to him. On the contrary, he would have restored the ancient Essene-Druidic sacraments. Christ is a Cosmic Power, the Chosen One of the Devas, the Solar Word, and not the dogma of organized religion, of Churches today. DOGMA reversed is AM GOD!

CHAPTER ELEVEN
The Unknown In Wyoming
By Mary Martin
VOL. 3 NO. 1 – October, 1981

Wyoming, or at least the northern part of the state, seems to be one of the most promising locations for an inner earth entrance. I have lived in Wyoming and can vouch for it being a very unusual state, with many mysteries attached to it, but for this article we will just touch on the highlights that concern our search.

MEDICINE WHEEL – To those who seek to restore the holy places to their rightful aspect, none is more sacred than the mysterious and brooding Medicine Wheel, It lies high in the Big Horn Mountains of Wyoming, at an elevation of 9,640 feet above sea level. Its age is unknown, as are its builders. The local Native Americans deny any knowledge of its construction, and a good many of them deny knowledge of its uses. The Medicine Wheel is an almost perfect circle, seventy feet in diameter. It is composed of 28 'spokes,' six of which terminate in cairns. The central 'hub' is twelve feet in diameter, and is surrounded by a stone wall two and one-half feet thick. This stone wall opens into a central circle, seven feet in diameter. At the perimeter of the wheel, the cairns are 'throne-like' in appearance. Five of these cairns face toward the

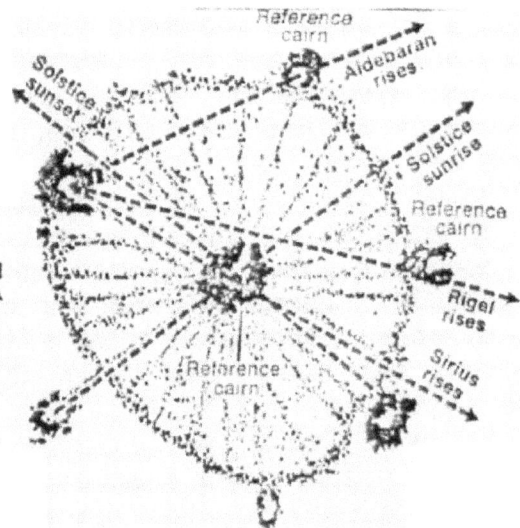

center of the wheel, while the sixth, the eastern, faces to the east. When the wheel was first discovered by the white man in the late 1800's, there were many Buffalo skulls placed around this eastern cairn. These are, of course, long since gone, as well as some of the stones comprising the wheel itself, due to vandalism. Archaeologists, some if who place the date of the Medicine Wheel at 13,000 years before the present, are mystified as to its use. Recently, an astronomer by the name of Dr. John Eddy, has offered an explanation of its purpose. He feels it is an ancient astronomical observatory. By sighting along the centers of two cairns to a point on the horizon, the ancient builders could accurately determine the rising of the sun at Summer Solstice. Others of the precise alignments indicate the rising of the stars, Higel, Sirius and Aldebaran. The twenty-eight spokes indicate the use of a Lunar Calendar.

"The alignments are so precise," Dr. Eddy is quoted, "that I don't think there's any question about what the Medicine Wheel is."

On the outside of the wheel, at distances from 70 to 275 feet are other monuments, all built on high points of ground. Indian legends tell of an underground cavern beneath the wheel in which the Alumdumbay, or "Little People" lived. These were very wise people, and great chiefs and medicine men went to this cave to learn wisdom. Wilford South, in a letter to me before his death, stated the following about Medicine Wheel: "Medicine Wheel was the Venusian base in earlier times; I would venture to remark that this mountain site could well be a centre cone leading to an underground city."

From the Hefferlin Manuscript, printed by the BSRF Foundation, we get the following regarding a tunnel in Wyoming:

"Another branch line ends in northwestern Wyoming, due west of Sheridan and some two hundred feet or more up the side of a mountain. This tunnel seems to have been twisted and sheared off, leaving a distorted and pinched outlet. When we consider the great density and toughness of the metal lining the tunnels, a metal that even earthquakes and great land mass movements cannot break, we wonder what titanic force sheared and twisted the Sheridan tunnel end."

EXPLORERS DESCRIBE HUGE CAVE IN WEST

Sheridan, Wyo (UPI) - A group of explorers have emerged from a cave in Northern Wyoming's Big Horn Mountains, saying it may be the deepest ever found in the nation. The team of six men

and one woman, all from the Sheridan area and ranging in age from 23 to 40, spent two weeks in what they dubbed the Great Expectations Cave, squeezing through narrow passages, swimming through pools over their heads and sleeping in darkness so black they often wondered whether their eyes were open or shut. They reached an estimated 900 feet depth before stopping... The country's deepest cave - Neff's Cave in Utah - is less than 1,200 ft. deep. ... The group reached a barrier they couldn't overcome without other equipment, but he declined to say what the barrier was. "We don't want to tell anybody that,"... "Right now we're one jump ahead of some other people and we don't want to give it away."
SOURCE: *ROCKY MT. NEWS* - 3/25/80.

TONGUE RIVER CAVE

The canyon looks insignificantly different than so many others in the state. Sharply-cut walls jut almost vertically from the shores of a raging mountain river that plummets down from the high country. But far below the familiar terrestrial features of the Tongue River Canyon lies another world where the glimmering sunlight and Prussian-blue sky never penetrate and darkness reigns supreme.

Spiraling down into the depths of the earth from a perch 400 feet above the cascading river is an enormous limestone cavern extending 6,700 feet into the subterranean world. Here in the nocturnal recesses lies an extensive network of tunnels, crawlways and rooms varying from tight constrictions barely allowing a person to crawl through, to an immense amphitheater known as the Boulder Room, Twisting down from a sloping room at the entrance of the cave, the main passageway leads to a long, high-ceilinged hallway known as Rainroom Number One, so named because of the continual dripping of underground water.

Rainroom No. 1 is the first of two such rooms encountered in the cave. Beyond this, through a long spiraling tunnel known as the Corkscrew, and past the Boulder Room flows an underground stream, which in 1961 was discovered to be an underground portion of the north fork of the Little Tongue River. Passageways are found both up and downstream, with the upper tunnel leading to a small waterfall and eventually disappearing into a siphon. Downstream a waterfall plunges 24- feet down to the cave floor. Unlike many areas of the U.S., Wyoming has only a few caves where diving is possible, though the Tongue River Cave is one such spot where it is thought additional caverns may be found by diving into the siphon where the stream enters the cave.

THE BEST OF THE HOLLOW HASSLE

It has long been thought by many people that the Sphinx marks the sight of an entrance to the underworld. The photo to the right is a sphinx that is found in the Bucegi Mountains, but the *Denver Post's Empire magazine*, Mar. 14, 1976 issue, carries a photo of a very similar formation near Laramie, Wyoming, I also found a sphinx, in another area of Wyoming. The location is not being given out at this time, but you can see for yourself in this photo at the bottom of the chapter.

LET US ALL CONTINUE OUR SEARCH FOR THE SECRET PLACES
OF THE LION!

CREDIT: "NEARA" NEWSLETTER - SUMMER

ARDALA - OF THE FOUR ARMS AND THREE FACES. VOL. 3 - NO. 3

MAY 1, 1982

CREDIT: PALMER PUBLICATIONS

CHAPTER TWELVE
Subterranean Sasquatch In The Shawangunks
By Donald Farmer
VOL. 3 NO. 3 – May, 1982

It seems that there is a continuing migration of the Sasquatch from the west toward eastern areas of the nation. The eruption of Mount St. Helens may be a contributing factor but the surreptitious Sasquatch movement eastward was perceivable before that Kali-like emergence from Mother Earth* Logic strongly asserts that the usual domain of the super-elusive Sasquatch must be subterranean. Thus, for a long time they might have been intimately aware of the seismic stresses steadily mounting toward a major earthquake on the west coast.

Or. instead of indicating flight from impending disaster, the Bigfoot wanderings could be motivated by the same quest we are concerned with... to find an accessible passage through the world-wide thermal barrier that may have been generated by limited but cataclysmal shifting of Earth's entire crust upon the inner solid shell, thereby effectively isolating our outer sphere from Inner Earth. A passage through this molten terrestrial divide is more likely to be located in a region of relative geological stability such as the eastern U.S.

If they are already exploring widely within near-surface Earth, the Sasquatch should be well qualified to assist us down that long, more difficult way to the core of the planet. Assuming we can somehow establish communication and co-operate with them, our technical abilities that took us to the moon would also be a great help to the Sasquatch on such a united trek to Intraterra.

A not so very remote region in which it might be possible to contact the Sasquatch is the Shawangunk (pronounced Sha-wan-GAN)

Mountains, often thought of as foothills of the Catskills, that extend southwest from the city of Kingston, N.Y. on the Hudson River. In an adjoining valley west of this ridge, U.S. route 209 passes through the town of Ellenville in southern Ulster county, N.Y, about 90 miles north of New York City. At the eastern edge of this town there is a tunnel of unknown origin in the base of the mountain. Back around the turn of the century it was quite a tourist attraction. There was a narrow railway with small cars that carried people to the tunnel's end where they drank from a spring of water claimed to be one of the very purest known. A reward of $1,000, substantial for the year 1900, was offered to anyone who could document the origin of this tunnel, but it remained uncollected.

Apparently the tunnel had been excavated by the use of hard metal tools which early Native Americans were not likely to have had. So a legend derived from another legend arose that ponce de Leon, during his search for the Fountain of Youth in Florida, had sent an expedition far northward and it was their Spanish steel that had dug this tunnel in an attempt to reach the source of the remarkable water. Hence, it came to be known as the Old Spanish Tunnel. Later it was closed and rumor suggested the Saratoga Springs resort bought it to eliminate the competition. The preceding history is from a book entitled "Treasure Tales of the Shawangunks," that has been misplaced, and it is hoped my memory is reasonably accurate.

Long before obtaining that book, when I was a teen-ager living at Ellenville temporarily in the mid-1940's the fabulous tunnel appeared merely to be a dank mess that could be entered only by squeezing past rotting timbers and getting wet feet. But upon my only return there a few years ago the tunnel entrance was closed and so completely concealed that I could not be certain exactly where it had been. Like too many other things, the Old Spanish Tunnel had vanished from the face of the Earth.

However, the so-called Ice Caves on top of that mountain, from which the purified water may originate, are probably more relevant to our main interest. There used to be a somewhat steep trail leading up to them from the valley. Mostly they are not walk-in or crawl-in type caves but several deep crevices, presumed to be of ancient glacial origin, that retain ice the year-round. It is of course obvious that nobody should attempt a descent into these depths without proper equipment and plenty of know-how. The 1981 Rand McNally Road Atlas indicates Ice Caves Mountain to be three miles southeast of Ellenville.

This is the general region where there have been many close encounters with Bigfoot referred to in the news item from the August 26, 1981 *New York Post*.

"BIGFOOT" STEPS INTO NEW YORK

The legendary "Bigfoot" - a lumbering, shaggy beast up to 10 feet tall - has been spotted in New York. The elusive, mythical creature - supposedly seen by many people on the mountainous northern Pacific Coast - Is now being reported in the forests north of the Big Apple.

A group of scientists whose hobby is investigating such phenomena say there have been 94 Bigfoot sightings reported In the region over the past 10 years; "As a scientist, I have to take the attitude that these things don't exist. But the conformity of

Sightings Is outstanding," said Robert Jones, an IBM computer analyst.

"We've got a variety of people in a variety of circumstances at a variety of times - and they ail say they saw the same thing."

Bigfoot's range In New York seems to be concentrated in an area south of Kingston and north of Poughkeepsie, but the beast has also been seen In New Jersey and Pennsylvania.

CHAPTER THIRTEEN
The Emergence Cycle
By Bruce Walton (Branton)
VOL. 2 NO. 3 – May, 1981

Since the dawn of time, it is said; mankind has undertaken great migrations to and from the Subterranean Worlds. Some say these migrations reoccur in cycles. The idea that life and time consists of cycles is one that can be found in such ancient cultures as the Egyptians, Incas, and the Hindus, as well as the disappearing Native American-Indian cultures.

Issue #3 of "The Source," published by Christine Hayes of Cortes, Colorado, carries the following passage: "To the Indian, time is a circle, the past is not irrevocably past, it is eternally present. What the Indian calls beginning time (or legend or sacred time) is not at the end of a long line stretching back into the past. It is not linear at all. It folds back on itself. Further, chronological timer in the generalized view, is a trap that results in loss of purity and potency. To get out of touch with beginning time is to become ill or unbalanced in mind, body or spirit."

Issue #1 of the same Newsletter continues: "Throughout the Earth's tumultous history. Cultures have entered and surfaced upon the planet in waves of human, tide. These migrators left the sediment of their impressions on the Surface World like the mineral rings in a tree, leaving-behind seed for new civilizations, taking with them' the primary knowledge of mankind. The last major migration was during the final destruction of Atlantis as she plunged her fiery heart into the fathomless well of the Atlantic. From this Great Migration many paths were taken, several of which ultimately led into the Inner-Earth domain.

"There are several Colony Cities located In the caverned regions between the surface and Inner Domain on both sides of the 'Skull Plate.' These colonies were formed from displaced Surface Dwellers in the many migrational surges to the core. For differing reasons, groups of migrators never Journeyed as far as the Central Cavity. These Colony Cities are still a part of the Central Earth Homogony."

Stories of the Great Migrations and histories telling of incredible journeys to and from mysterious-subterranean lands have been recorded on every continent of the world in the form of beautiful oral traditions and legends.

Edward Bulwer Lytton, popular author of "The Last Days of Pompeii," published a lesser-known novel which turned out to be one of his most unusual and controversial stories. Because of his reputation as a fiction writer, this story, titled "THE COMING RACE" was immediately labeled as another one of his works of fiction. The German Nazis had other ideas however. The LUMINOUS LODGE OF THE VRIL SOCIETY was their answer to Lytton's story. Its major purpose being to seek out these underground beings mentioned in his book and bring back their "VRIL" Power, an energy source of tremendous potential which could be used as an instrument of great good or evil. The Nazis were determined to get hold of this power source by any means possible so that it could be used by Hitler as a weapon; with which he would use to conquer the world.

The anonymous informant, who is alleged to have given Lytton the story, never revealed the location of the entrance to the subterranean lands, of the "Ana," as the informant calls them and needless to say, the Vril Society failed, and the "VRIL," if it exists, still remains in the hands of a subterranean civilization.

"THE COMING RACE" is the story of a young miner who, while exploring a strange chasm that he had broken into, entered into a strange underground world where he found a race of humanoids far in advance of surface nations, a race who called themselves the "Ana." He stayed with these people for quite some time, learning their language, culture, science and religion. One of the Ana, a subterranean woman named "ZEE" told him of their ancient tradition of how their race came to live in this underground land. She told him that their nation had at one time in the far distant past dwelt on the surface of the Earth. At that time the remote progenitors of the race had been subject to many, violent revolutions of nature. Cataclysms on a great scale had caused their homeland to become flooded.

"...A band of the ill-fated race, thus invaded by the Flood, had, during the march of the waters, taken refuge in caverns amidst the loftier rocks, and, wandering through these hollows, they lost sight of the upper world forever. Indeed, the whole face of the earth had been changed by this great revulsion; land had been turned into sea, sea into land. In the bowels of the Inner Earth even now, I was informed as a positive fact, might be discovered the remains of human habitation, habitation not In huts and caverns, but in vast cities whose ruins attest the civilization of races which flourished before the age of Noah, and are not to be classified with those gerera to which philosophy ascribes the use of flint and ignorance of Iron..."

The significance of the books title becomes apparent with the following words which were spoken by ZEE to the anonymous adventurer.

"...With our race, therefore, even before the discovery of Vril, only the highest organizations were preserved and there is among our ancient books of legend, once popularly believed, that we were driven from a region that seems to denote the world you come from, in order to perfect, our condition and attain to the purest elimination of our species by the severity of the struggles our forefathers underwent, and that, when our education- shall become finally completed, we are destined to return to the upper world, and supplant all the inferior races now existing therein."

The Incas, most prosperous of all recent western cultures, until invaded and conquered by the Spaniards, also believed in a great cycle. There are reports that the Incas actually escaped into tunnels when their nation was invaded, and they, are now believed to live in subterranean lands beneath the Mato Grouso district of Brazil, with a race of Atlantean survivors of the Great Atlantean cataclysm of 12,000 years ago. Local Indians, descendants of the original Incas, have a legend which states that the first Incas emerged front three openings in a cliff about 20 miles from Cuzco. It should be stated here, however, that there are other traditions of the first Incas having their origin on Atlantis, or even from "the sky," and these differing legends make it hard to determine their exact point of origin. The South American Indians also believe that their ancestors, the Incas, will one day return to the surface of the Earth.

"...One day, say the Indians, the wheel of life, or cycle of events, will come full circle, and the ancient people will return and reintroduce a golden age. (As one has seen, millennial prophecies of this kind are common all over the regions where the Atlantean Central and South American empire once held sway.)" —

("MYSTERIES OF ANCIENT SOUTH AMERICA" by Harold T. Wilkins.)

The Lost Tribes of Israel, who are said to be living in a hidden land in the extreme north, according to the Bible, will one day return and become reunited with the tribes of Judah and Bengamin (Jews). The account of the arrival of the Ten Tribes into the land of the far north can be found in chapter 13 of the Book of II Esdras in the APOCRYPHA.

"...And whereas thou sawest that he gathered another peaceable multitude unto him; those are the ten tribes, which were carried away prisoners out of their own land in the time of Osea the king whom Salmanaser the king of Assyria led away captive, and he carried them over the waters and so came they into another land. But they took this council among themselves that they would leave the multitude of the heathens and go forth into a further country, where never mankind dwelt that they might keep their statutes, which they never kept in their own land. And they entered into Euphrates by the narrow passages of the river for the most High then shewed for them, and held still the flood, till they were passed over. For through that country there was a great way to go, namely, of a year and a Half, and the same region is called Arsareth. Then dwell they there until the later time, and now when they shall begin to come, the Highest shall stay the springs of the stream again, that they might go through, therefor sawest though the multitude with peace..."

The return of the Lost Tribes of Israel will be one of the greatest periods in Earth's entire history, as we can see by reading "EVERYMANS TALMUD" by Dr. A. Cohen, page 354: "...Another confirmed belief was that the Messiah would effect the reunion of the tribes of Israel. While we find the teaching 'the ten tribes will have no share of the World to Come.' (Tosifta Sanh XIII. 12), the Talmud usually takes the opposite view. By appealing to such texts as Is.xxvii.13 and Jer.iii.12, the Rabbis enunciated the doctrine of the return of the lost ten tribes (Sanh II ob:}. 'Great will be the day when the exiles of Israel will be reassembled as the day where heaven and earth were created.' (Pes. 88a). A law of nature will even be miraculously suspended to assist this great reunion. 'In the present world when the wind blows in the north it does not blow in the south, and vise versa; but in the Hereafter with reference to the gathering of the exiles of Israel, the Holy One, blessed be HE, said, I will bring a north-west wind Into the world which will affect both directions; as it is written: "I will say to the north, give up, and to the south, keep not back; bring; My sons from afar and my daughters from the end of the earth." (ls.xliii.6))' (Midrash to Esth.i.8)..."

Long ago, it is said, the inhabitants of the earth abandoned the surface in search of a new life deep underground. There they were able to evolve in technology and spirit, with no fear of invasion by 'outsiders.' safe from the cataclysmic changes that have so many, tines in the past altered the face of the earth, and free from diseases that afflict so many inhabitants of the Outer World.

One amazing parallel between these various stories of subterranean dwellers are the reoccurring prophecies telling of the return to the surface of those nations who abandoned the Outer World so many thousands of years ago. The following accounts are just a few that I have come across which suggest that our world is on the verge of another great cycle of emergence, spanning the period of transition of the Earth into the Aquarian Age.

"...A Polynesian legend describes the ancient race living deep beneath the ruins of the stone city on the South Pacific Island and says they will someday emerge to again rule the earth. A peculiarity of the construction of these buildings is the odd stone shapes which make the structures look somewhat like colonial forts..." ("SUBTERRANEAN WORLDS OF PLANET EARTH" by Gene Duplantier, p. 9)

There is a Mongolian tradition which states that over 60'000 years ago, a Holyman disappeared with a whole tribe of people under the ground and never appeared again on the surface of the earth. They became part of a great and powerful nation called the AGHARTI.

"...All the people there are protected against evil and crimes do not exist within its boundaries. Science has there developed calmly and nothing is threatened with destruction. The subterranean people have reached the highest knowledge. Now it is a large kingdom, millions of men with the King of the World as their ruler. He knows all the forces of the world and reads all the souls of humankind and the great book of their destiny. Invisibly he rules eight hundred million men on the surface of the earth and they will accomplish his every order." ("BEASTS, MEN AND GODS" by Ferdinand Ossendowskl, ch.XLVI-XLVIII)

In the Altai Mountains of western Mongolia, in the beautiful upland valley of Uimon, it is said there once lived and flourished the powerful tribe of Chud. "They knew how to prospect for minerals and how to reap the best harvest. Most peaceful and most industrious was this tribe. But then came a White Tzar with innumerable hordes of cruel warriors. The peaceful industrious Chud could not resist the assaults of the conquerors and not wishing to lose their liberty, they remained as serfs of the White Tzar. Then, for the first time, a white birch began to grow in this region. And, according to the old prophecies, the Chud knew that it was time for their departure. And the Chud, unwilling to remain subject to the White Tzar, departed under the earth. Only sometimes can you hear the holy people singing; how their bells ring out in their subterranean temples. But there shall come the glorious time of human purification and in those days the great Chud shall again appear In full glory." ("SHAMBALLA" by Nicholas Roerich, pp. 210-222)

The Tupari Indians, who live on the Rio Branco-Parima river In the upper Mato Grosso region of Brazil have the following tradition, which was recorded: by a young Swiss ethnologist, Franz Casper when he visited the tribe in 1948:

"Long ago there were no Tupari or other men. Our ancestors lived under the ground where the sun never shines... Then the men began to stream out in great hordes... Many men remained inside the earth. They are called 'Kinno' and still live there. One day, when all the people of the earth have died, the Kinno will come out of the ground here. But the men whom Aroteh had let out of the earth' did not find room in the same place, We Tupari remained here, the others wandered far away In all directions. They are our neighbors, the Arikapu, Yabuti, Mahurap, Arua, and all other tribes." ("SOUTH AMERICAN MYTHOLOGY" by Harold Osborne, P. 119).

The January 1947 Issue of "AMAZING STORIES" magazine carried the incredible story of a woman by the name of Margaret Rogers; the story put forward as a factual account tells of her three year stay with a tall race of highly advanced subterranean beings called the NEPHLI. During her stay, she had many amazing experiences, too numerous to fully describe here. Among the fantastic things she reportedly observed during her stay were telescopes that could penetrate the most distant galaxies, a machine that could "tune-In" on any part of the earth, and machines that could pick-up any radio or television stations.

She traveled in strange cars that were run by the power of thought and could reach speeds of almost 2000 miles per hour. She observed machines that could make the old young again, thus lengthening the life expectancy dramatically. She observed huge Space Ports and ships that constantly made journeys to distant, stars, and she was shown huge garden caverns where many strange fruits and vegetables were grown. She was shown the history of the NEPHLI on their eternal records, how they came to earth hundreds of thousands of years ago from the Mother Planet, and how they built beautiful cities on earth where they lived in peace.

But .then some of those who came to earth with the NEPHLI as colonists strayed from their teachings. The NEPHLI could have destroyed them like flies, but they were a peaceful people and wished no harm to cone upon their fellow man.   One of the subterranean beings named Arsi told Margaret: "...We are all human, though those on the surface would not call us so.   The NEPHLI civilization was far advanced when we went underground.   Those on the surface strayed from their teachings and scorned help from the Mother race. Now see to what they have come. How they are a proud and arrogant people who would have had more to be proud of if they had followed the teachings of their ancestors. Remember all of this when you return."

The entrance to this Subterranean World is reported to be somewhere near Ixtaccihuatl, Mexico, with another possible entrance in a cave called the "Cave de los Vientos."

When asking Arsi about the spacecraft she had observed, Margaret was told: "That is the energy-driven ship that surface men will someday use to go to the stars." He added sadly, "Up and up he will go, not alone to the stars, but to the other sciences as well, until his arrogance and pride leads him to believe that he can reach to Tamil (GOD) himself. Then, Maggie, Tamil in his wrath that man should try to assume the attributes of the supreme beings will destroy him and all his works and of the surface people leave only those who are humble and clean hearted. Shall I tell you what will happen then? The NEPHLI will come back to the surface, to their rightful heritage and bring all their marvelous science to make the world a peaceful place to live in, a world of beauty where wars are no more. Then, and only then, will Tamil be fully revealed to us."

Our world is even now being prepared for this greatest of all reunions between the lost tribes of man, but before this   long awaited reunion can come about, the Earth must pass through the purifying transition manifesting itself in the form of man-made and natural upheavals.

As the King: of the World in Agharti prophesied when he visited the Narabanchi monastery in the year 1890:

"...All the Earth will be emptied. God will turn away from it and over it there will only be night and death. When I shall send a people, now unknown, which shall tear out the weeds of madness and vice with a strong hand and will lead those who still remain faithful to the spirit of Man in the fight against evil. They will found a new life on the earth purified by the death of nations. In the fiftieth year only three great kingdoms will appear which will exist happily for seventy-one years. Afterwards, there will be eighteen years of war and destruction. Then the peoples of Agharti will come up from their subterranean caverns to the surface of the earth."

WHAT LIES BENEATH

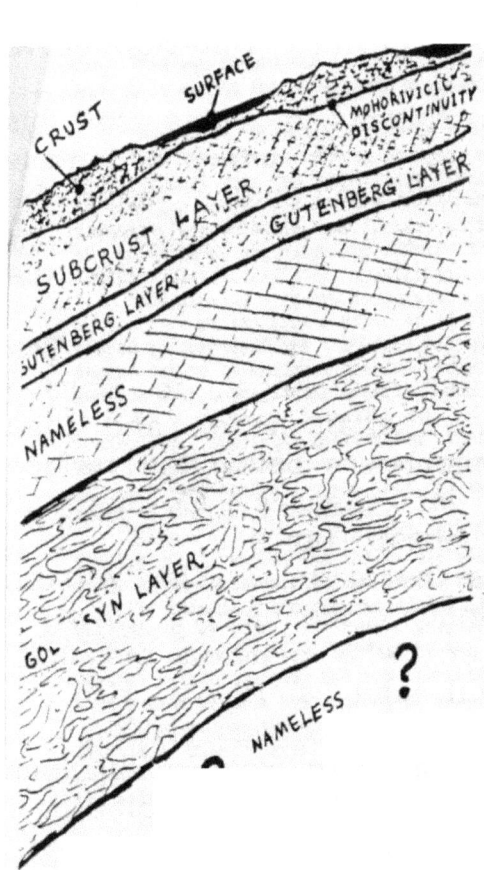

MOHOROVICIC DISCONTINUITY (MOHD) -
Division line separating the Crust and the Mantle of the Ekrth. Located at a depth of 10-60 km beneath the surface. The crust is thinnest beneath the oceans and thickest beneath the continents.

SUBCRUST LAYER - Magma is non-existent in this and lower layersf according to the latest discoveries by Russian scientists. Seismic velocity (the speed of which seismic waves travel through the Earth) increases upon entering this level.

GUTENBERG LAYER - layer of relative repose, contains little activity.

NAMELESS - This layer does not seem remarkable for anything in particular, and there is relatively little activity.

GOLITSYN LAYER - High levels of activity have been recorded in this layer, although scientists are at a loss to explain the reason for this.

NAMELESS - This zone is almost completely quiet, and has been traced to depths exceeding 2,900 km. Beginning here the seismic waves mysteriously reach an almost limit velocity of 12-12.5 km. per second. Is this favorable evidence of the existence of a hollow cavity approximately 800 km. beneath the surface?

This diagram is based on a description of the Earth's interior which appears in A. Malakhov's book *THE MYSTERY OF THE EARTH'S MANTLE* (Peace Publishers., 2, Pervy Rizhsky Pereulok, Moscow, U.S.S.R, - Translated from Russian by Ekvid Sobolev), pp.96-98. This excellent book describes the recent findings made by Russian scientists during their deep-drilling projects which were and are being carried out in the Kola Peninsula and other sites.

In the chapter titled "THERE'S NO MAGMA UNDER THE CRUST!" there appears the following information. Note: Those who believe that the Earth's interior is composed of molten rock and lava are in the scientific community referred to as Magmatists, while those who believed in the cold-interior theory are referred to as Neo-Neptunists. Although in the past the majority of the world's geologists considered themselves Magmatists, this is slowly changing as new evidence surfaces in support of the Neo-Neptunist's hypothesis

"There is no magma! Scientists began to come to this conclusion long before the deep hole was sunk into the famous Moho discontinuity. This conclusion became self-evident after the extraordinary phenomenon which was established and registered by instruments let down together with the drilling equipment into the depth's of the Earth's crust.

"Not much attention was attached to it at first. The temperature registered by special instruments began to behave strangely. First, the curve stopped at a certain level, and then began to drop Instead of rising. At 10 kilometers below the surface the temperature was already close to zero degrees Celsius Still deeper the temperature passed into the subzero range...

"This alarmed everybody. Where is the magma? What will be found when the hole reaches the Moho discontinuity? How to account for the mistake of the engineers who planned the unit? They designed the equipment to withstand a sharp rise in temperature as it sank further down. All their work was in vain.

"New equipment had to be made in a hurry, new calculations performed, and new materials sought for super-deep drill-holes in a zone of high pressure and negative temperatures.

"Meanwhile, incredible reports came in one after another...the temperature kept on dropping with depth.

"Report from the Temperature Front – that is what the newspapers called this unusual information, which was accompanied by other data just as curious. That is when Academician Vernadsky's hypothesis, put forward as far back as 1934, was recalled... Unfortunately, the scientists, engineers and constructors who equipped the drill-hole had not taken it into account...

"And finally a record temperature was reached. The most interesting analysis was that of the core raised from a depth of 15 kilometers on the Kola Peninsula. Here the hole had passed through half the thickness of the Earth1 s crust. The rock raised from this depth turned out to be ordinary granite, very similar in appearance to those that crop out on the surface over a considerable part of the territory of the Kola Peninsula and Karelia.

"But the greatest sensation was the age of these granites. Conventional methods were inadequate. Specially constructed atomic time counters registered an unheard-of age. Almost 100,000 million years! When designing this hole the scientists had expected to encounter some very ancient rocks. But not 100,000 million years. This refuted all the familiar ideas of the age of our planet.

"Pieces of the core from the Kola Peninsula were tested in all the geochemical laboratories of the world, and the result was the same everywhere. Hence, the scientists who spoke of the immeasurable age of the Earth were right, after all.

"By this time the results of drilling the other holes had come in. The Urals hole was just as Interesting. Here a strange kind of substance was raised from a great depth. At first scientists were baffled by the fact that over a comparatively large Interval of several kilometers they had not been able to raise a single piece of rock to the surface. At the same time, a sharp odor of gasoline spread all over the area round the drill-hole.

"Then one of the drill foremen suggested that special devices should be designed capable of raising greatly compressed matter from the depths. A soil pump was made, and a dense black mass was raised to the surface. It was petroleum. Petroleum compressed to the solid state.

"Hence, it was worth while prospecting for deep-seated oil in the Urals. Is petroleum, therefore, a product of the depths of the Earth?

"Still more striking news came from the super-deep drill-hole in the Kuril Islands district. There the rocks of the Moho frontier were struck for the first time. The hole sank into a substance which was under a very high pressure, as expected, but at a low temperature. When the substance was raised to the surface it was found to be an ordinary piece of compressed rock of the basalt type. It contained a large amount of different metals copper, rare elements and other metals. The Kuril hole came across no metals.

"The neoneptunists were triumphant. This confirmed their conclusions which for a long time had been considered doubtful. The world press was full of data which irrevocably defeated the magmatic theory of ore formation and mountain building.

"Now everything was in order. Textbooks were quickly revised so they no longer gave hypothetical ideas, but actual facts. Projects appeared for drilling deep and super-deep shafts. Such shafts would start with an immense diameter of several scores or meters, so that they could gradually narrow to the conventional size.

THE BEST OF THE HOLLOW HASSLE

"The revolution in research techniques ushered in by the first five exploration holes opened out new horizons. There was always a long queue at the Academy of Sciences Building at the National Economical Achievements Exhibition. Here a diamond of the first water was on display, the size of a man. This was a synthetic diamond obtained on the basis of the data obtained from drilling the super-deep hole that reached down into the lower temperature zone. It appeared that diamonds of enormous size formed comparatively easily under those conditions!

"So that the visitors would have no doubts that the diamond was of artificial origin, the words: "Peace, Work, Freedom" were laid out inside it in a combination of colored minerals..."

"Finally, a core was raised from the zone of the Moho discontinuity, a piece of it was sent to the biological laboratory, where quite unexpectedly traces of life were discovered in it. This incredible find excited scientists even more than the hard rock and gigantic diamonds. Life does not confine itself to the thin film near the surface of the Earth. There is life at such inaccessible depths too.

"Long before this hole had been drilled the soviet scientist T. L Ginsburg-Karagicheva detected bacteria at a depth of three kilometers in one of the wells of the Apsheron Peninsula. These bacteria differed from those on the surface in that they easily adapted their vital activities to life without access of oxygen. But the bacteria brought up from the super-deep holes assimilated oxygen. They were more highly organized than those discovered by Ginsburg-Karagicheva. They obtained oxygen by decomposing rocks into their constituent parts. They were a special kind of oxygen bacteria.

"And again the question arose before the scientists! How did life originate on Earth? Again numerous hypotheses were put forward, consisting essentially in the statement that life had broken out on to the surface from the depths of our planet. But how had it come into being there? That gave the scientists something to think about.

BEYOND THE MOHO DISCONTINUITY

"Late at night when only the night shift of drilling engineers remained at the drill-hole on the Kuril Islands a strange thing happened, which set the whole town of drillers into A flurry. Alarms were heard on all sides. The whole town, all of its inhabitants gathered at the drill-hole.

"And the drill-hole really was a sight. It was all aglow with a weird bright light. Tongues of cold flame shot out of people's fingertips, then jumped to their hair. The derrick was buzzing and humming strangely. A strong smell of ozone permeated the air.

"This glow phenomenon has been known to man since antiquity. In ancient Rome it was known as the Fire of Castor and Pollux. In the 16th-17th centuries it was named after saints in whose honor various churches had been built. The glow was seen mainly on sharp objects in high places, and that is why people saw it most frequently on the cross at the top of a church. At one time it was called the Fire of Saint Erasmus; later it became known as Saint Elmo's Fire. It was this "fire" that had alarmed the drillers* settlement on the Kuril Islands.

"The Elmo Fire arises, as a rule, when there is a high electric field in the atmosphere. Sometimes it is very high-as much as 30 thousand volts per centimeter.

"Measuring Instruments were quickly put into action. It turned out in this case the electric field was unusually high, running into millions of volts per centimeter.

"Day after day the electric field recorded in the logbook rose higher and higher. The deeper the drilling equipment sank into the Earth, the higher it rose.

"Experts from all countries were unable to find the reason for this phenomenon. Only gradually, as facts accumulated, did the answer begin to come to light. The investigators of the depths had come across a constant flow of electricity, the source of which is situated in the zone near the Earth's mantle.

"The stray currents observed by scientists in the zone could not be compared with what was encountered near the Moho frontier. So the scientists who had put forward the idea of an Earth dynamo were right! This was a confirmation of their idea that the Earth is an immense electromagnet. But even those scientists had not thought that the electric field would be encountered so close to the Birth's surface, just below the crust. They said that the Earth dynamo is related to the electric current moving in the zone close to the Earth's core, underneath the mantle. And it suddenly turned out that there is an Earth dynamo near the Mohorovlcic discontinuity as well.

Subsequent investigations of the electric field led to interesting conclusions. The value of the electric field varied depending on the processes taking place on the Sun. Scientists established short-period variations near the Moho discontinuity, related to changes in the life of the Sun. They found dally, monthly and yearly variations. This confirmed the idea suggested by certain scientists that the life

THE BEST OF THE HOLLOW HASSLE

of the Sun and that of our Earth are related by common processes. Projects for utilizing the electric field in the Earth's bowels began to spring up in different countries. One of the significant ones was that of engineer Ivanov, who suggested building a Global Power Grid by drilling a series of super-deep wells. One of them should be sunk in the district of Prince of Vales Land, where the northern magnetic pole is situated, and another in Antarctica, at the southern magnetic pole. A number of intermediate wells would then have to be drilled in various parts of the globe, and they could be connected up into an integral power grid.

"Mankind had received a free source of electricity of unheard-of power..."

CHAPTER FOURTEEN
Quest For The Lost Kingdom of MU
By Bruce Walton (Branton)
VOL. 5 NO. 3 – May, 1984

Several years ago Sebastian Balfe Dangerfield, one-time Associate Curator of the Museum of Archaeology in New York City related to a few of his close friends a fascinating discovery which was Bade by Christian Thornton of the Kon-Tiki expedition a discovery which was not revealed to the public.

Sebastian, who was in regular correspondence with Thornton, was told that while the members of the expedition were in the Jungles of Peru searching for balsa trees they came across a hairy hermit, who spoke in the garbled tongue of the natives a tale of gold, fair women and a city where the Gods play. Here Thornton came in contact with this same dirty old hairy hermit and through certain rare relics found upon his loathsome body they came to the conclusion that he was telling the truth.

Sebastian himself later Journeyed to Peru, with the support of the National Geographic Society, to find out "if this is the lead we have been waiting for or just another old dirty hairy hermit hoax."

Sebastian later found himself in the Jungles of Peru, and confirmed the existence of the relics and the hermit, who consented to guide them to the "City of the Gods."

"I know from the relics," adds Sebastian, "from the faint inscriptions on them that we have labored into English and from the hermit, this hairy hermit whose eyes light up in visions of splendor that is beyond his garbled tongue, but in his manner certain gestures and a look about him that he unleashes in the night, squatting by the campfire and scratching antic figures in the dust...he has been there, and God willing, I shall follow him to the very seat of an his wildest dreams to that lost but happy Kingdom of MU."

Details on the outcome of this expedition are mysteriously lacking, other than the fact that the expedition was wrought with much difficulty, injury and misfortune. Whether or not they finally reached their goal is not known, but Christian Thornton did, according to reports discover a cave near the walls of a high range of mountains, which contained inscriptions "very similar to cuneiform." Here, then, is Thornton's translation of the mysterious message left ages ago by an unknown adventurer:

THORNTON'S TRANSLATION OF THE INSCRIPTION ON THE CAVERN WALL:

"O WHAT CAN AIL THEE, KNIGHT AT ARMS, ALONE AND PALEY LOITERING, THE SEDGE IS WITHERED FROM THE LAKE AND NO BIRDS SING."

TO THE GODS DO I LEAVE THE SUNLIGHT, FILLING MY LUNGS WITH AIR. FOR INTO THIS DARK CAVERN I MUST ENTER AS INTO A TOMB WITH LITTLE HOPE AND AN OVERPOWERING FEAR THAT I AM ALREADY TOO LATE. FROM THE FOREST I TRACED THEIR FOOTSTEPS READING HIS CLEAVED SANDELS HEAVY IN THE MUD AND THE FAINT BIRD-TOES BEHIND HIM AS IF SHE RAN BLINDLY AND SHORT OF BREATH, EAGER TO SURPRISE HIM LIKE A COLT BEHIND ITS MOTHER GAMBOLING FROM SIDE TO SIDE YET QUICK TO FOLLOW WHENEVER THE DISTANCE GREW. AND DO YOU GRIN, FAT MAN, MUSTACHED WITH YDUR CHESHIRE SMILE? O, BY THE SAINTS, I SEE YOU SWEATING IN THE SANDS AS YOU BEGIN THE ASCENT. SADLY I SIFT THE STORY FROM THE DUST. HERE IS WHERE YOU TOOK HER HAND, PERHAPS TO GUIDE HER UP THAT HUGE CREVACE IN THE BOULDER. AND DID IT FLUTTER LIKE A BIRD SLIGHTING ON A LIMB FEARFUL OF ITS TRUST, OR DID IT PINION YOU SECURELY AS IF TO ANCHOR TO A ROCK AND HERE, WHERE THE STORY ENDS IN THE DUST, RED DUST AS IF THAT GAPING HOLE WERE BUT A TOOTHLESS MOUTH AND THIS ITS LIVING TONGUE. HERE I WATCHED HER FOOTSTEPS AS THEY QUICKEN WITH ANTICIPATING JOY, AND THEN HER HESITATION. HOW QUICKLY YOU OVERCOME IT. SEE HOW SURELY SHE SLIDES. WHY, WHAT IS THE PROTEST OF HER HEELS TO THE STRENGTH OF YOUR ARMS; ONLY TWO NARROW FURROWS QUITE SMALL IN THE LIVID DUST AND ALMOST ERASED BY THE WIND. AND ARE YOU STRONG? O, MAY THE DEVILS MAKE YOU STRONG.

HEALTH TO YOUR ARMS, THAT I MIGHT FIND YOU WELL TO FEEL YOU STRUGGLE WHEN MY FINGERS DAGGER IN YOUR THROAT. O, GOD, IF THEY HAVE HURT HER, GIVE ME THE POWER TO KILL. FINGER ME IN TIGER-CLAWS AND FANG ME SHARP AS THE CROCODILE WITH THE MIGHTY ARMS OF THE GIANT ANTEATER. SEND ME TO SLAUGHTER.

O, MY LOVE, MY LIFE, WHY DID YOU FOLLOW HIM? WITH VOICE SO SOFT AS IF IT HAD CHAMBERS WITHIN ITSELF FILLED WITH SWEET ECHOES REPEATING SOUNDS THAT PLEASURED ME FROM SOME FORGOTTEN AGE AS IF THIS SLAVE WERE ONCE HER LOVER AND LAUGHED AND GREW WITH HER IN KINGDOMS TIME FORGOT. AND IN THOSE GLANCES OF HER EYES PRIDE AS IF ONE FLUNG A TORCH HIGH INTO THE NIGHT AND WHILE IT FLEW I HELD MY BREATH IN WORRY FOR IT; AND BEAUTY AS IF THOSE EYES HAD SPUN IN ORBIT WITH THE STARS WHERE THE GODS CAUGHT THEM IN A GOLDEN NET AND TRACED AN IMAGE OF ALL THE PLAY OF THE HEAVENS IN THEM BEFORE THEY CAST THEM BACK SPINNING IN HER HEAD. EYES THAT I TRACE MY CHILDHOOD IN AND WITH A SPACE AS IF THE UNIVERSE WERE NEW AND THERE BEFORE YOU WAS A SILVER PATH STRAIGHT AS THE ROAD TO GLORY.

O, MY LOVE, I MUST FOLLOW THEE EVEN INTO HELL. INTO THE HANDS O LORD I TRUST MY SOUL AND FATE. I HAVE HEARD OF THE DEVILS AND THE TROGLODYTES DEEP IN THE CAVERNS. I FEAR WITH ALL MY BEING, BUT LORD IF THOU CANST NOT HELP HE I MUST STILL GO ON FOR THERE IS NO LIFE WITHOUT HER. AND IF I DIE AND YOUR SPIRIT BE FREE THEN COME AND TRACE ME ON THE ROCKS THAT I MIGHT WARN ALL ELSE WHO ENTER HERE.

CHAPTER FIFTEEN
In Quest Of The Subterranean World: My South American Adventure in Search of the Unknown
By Raymond Bernard, A.B., M.A., Ph.D.

It was back in 1950's when I first put foot on Brazil, as I landed by plane in Belem, the great port near the mouth of the Amazon that I first learned about the Subterranean World, except for references to Agharta in Ossendowski's "Beasts, Men and Gods," the famous novels "Etidorhpa" and Bulwer Lytton's "The Coming Race" and references to the tunnels of the Atlanteans under South America in Harold Wilkins1 books "Mysteries of Ancient South America" and "Lost Cities of Old South America", and also Nicholas Roerich's "Heart of Asia."

Walking along the streets of Belem my eye was attracted by a large building in front of which were letters in cement, "BRAZILIAN THEOSOPHICAL SOCIETY." Curious to know about the progress of the Theosophical movement in Brazil, I entered and met a Mr. de Souza, the head of the society. He said this was only a branch of the central society with headquarters in Sao Lourenzoa in the state of Minas Gerais, where lives Professor de Souza, who was its head.

Though my knowledge of Portuguese was quite meager, as soon as Mr. de Souza began speaking I noticed that his main subject of conversation was "the Subterranean World" and the tunnels that run under Brazil and lead to subterranean cities. In these cities, he said, live Advanced Beings - descendants of the ancient Atlanteans. It was very fascinating and when I asked for more information he told me to visit Professor de Souza, a great archeologist, who is the authority on the Subterranean World.

Years passed and in 1957 when browsing in a bookshop I happened by a paper-bound book bearing the intriguing title, "FLYING SAUCERS FROM THE SUBTERRANEAN WORLD TO THE SKY." On the cover was a picture of a flying saucer leaving the underworld and going up into the sky. The author, O.C. Huguenin, claimed that all flying saucers come from inside the earth and that none come from other planets, which idea, he claimed, was false. And I noticed that the book was dedicated to Professor Henrique Jose de Souza and his wife, Helena Jefferson de Souza.

At that time I believed that flying saucers came from other planets, and the idea that they came from inside the earth seemed strange and impossible. I did not know then about the earth being hollow with an opening at the North Pole through which saucers enter and leave, and it seemed incredible that they could come from openings in the ground, though there are certain places, as in Antarctica and near Salta, Argentina, where saucers have been observed to ascend and descend from a common spot, indicating some base or subterranean entrance there.

I then remembered that the Theosophical leader in Belem had told me that Professor de Souza resided in Sao Lourenzo, so anxious to learn more on the subject of the subterranean origin of the flying saucers, I at once phoned him and he invited me to visit him.

On arrival in Sao Lourenzo, a mineral hot springs bathing resort, I was met at the station by a delegation of people who spoke English and seemed to have come from many nations. They said they were members of Professor de Souza who have come and made their home in Sao Lourenzo, center of the Brazilian Theosophical Society of which they were members. They then drove me to an immense temple in Greek style over whose entrance I saw the word "AGHARTA." That reminded me of the reference to "Agharti" in Ossendowskil s book mentioned above - it being the Buddhist name of the Subterranean World.

After showing me around the temple they brought me to a museum room. In it were kept various articles which they claim were brought up from the Subterranean World. One of these was a large glass jug of water, covered with woven straw. They all seemed convinced the various articles in the museum came from the Subterranean World.

They brought me to a hotel, and the next morning they came early to bring me to the house of Professor de Souza, who was by the casual waiting for me. He said that he had returned from one of his visits to the Subterranean World, where he was well known, and that he once had in his possession the key to the door that leads to Shamballah.

Shamballah? Where did I hear that word before? I read about it in Ossendowski's book and also in Roerich's work on his travels in the Far East, where it is believed that Shamballah is the capital of the Subterranean World of Agharti or Agharta, where resides the king of the world in his golden palace. In fact Shamballah, the heavenly city, is the central object of reverence of millians of Buddhists in Mongolia and Tibet, who say, "Hail Shamballah, thou Champion of Aghartal"

I noticed on a sofa in the rear of the room a young lady sitting, who seemed to be about 18 years of age. I thought it was Professor de Souza's daughter. But she was introduced to me as Helena Jefferson de Souza, wife of Professor de Souza, who seemed to be about 70 years of age. Knowing that I would be puzzled, it was explained to me that she is not of this world at all, but comes from the Subterranean World, where people never grow old, and is really over fifty.

The Professor then began talking about Colonel Fawcett, who disappeared in the jungles of Mato Grosso a quarter of a century ago, while in quest of the lost cities of the Atlanteans. The Professor said that he and his son Jack are both living in the Subterranean World and were not killed by the Chavantes Indians as commonly believed. When last seen he was heading to the Roncador Mountains of northwest Ma to Grosso, after leaving the city of Culaba. It was a month's journey through impenetrable jungle inhabited by fierce Chavantes Indians who act as guardians of the tunnel opening in Roncador that leads to the still inhabited city of the Atlanteans for which he sought.

I then asked the Professor to tell me how I can enter the Subterranean World as I wished to visit it. He told me to go to Roncador and follow Fawcett's footsteps, and to protect myself from the poisoned arrows of the Indians, he gave me a password and said that if I shouted this password when I came near the Indians, they would not molest me and let me pass safely. He then gave me a letter of introduction to a member of his society in Cuiaba, who would help organize an expedition to Roncador; and I was soon off on a plane to Cuiaba, chief city of north central Mato Grosso.

The word "Mato Grosso" in Portuguese means Great Forest, and so it is. The hinterland is the least settled area of Brazil and contains the world's biggest forest, being part of the Amazon basin which contains about one-fourth of all the timber in the world. For hours the plane flew over uninhabited forest; and finally I reached Cuiaba.

The elder people of Cuiaba still remembered the "Englishman" who came to their city some years ago, stayed a while and then went off into the jungles heading toward Roncador. A blond Indian boy, the son of Jack Fawcett, was living there then, born of an Indian mother. I met a missionary who said he was stationed at the last outpost that Fawcett left before he entered the jungle, heading northeast in the direction of Roncador. He claimed that later, his three Indian guides returned to confess that they killed Fawcett because they refused to go on in spite of Fawcett's insistence that the party proceed on a month's journey through jungle infested with fierce Indians. They said that their only recourse was to kill him and so be able to return. However, the story of the missionary sounded suspicious, for it was very unlikely that the Indian guides would confess having killed Fawcett even if they had done so.

However, I was determined to go on and reach Roncador, but could not undertake the journey alone and did not know the route. However, that problem was solved when one day I met in a cafe an old black man, who was somewhat drunk and talked freely. He said he was the son of Fawcett's chief guide, who brought him to the land of the "White Indians" who live in the general area of Roncador and Bananal Island. He said that he had spent most of his life among these Indians, whom he described as a race of fruit eaters and who were of fair complexion and much different from the surrounding darker races that eat meat and are more savage.

For these white Indians, the colored man said, were not savages, but highly cultured and knew how to write, for he saw among them strange inscriptions in an unknown language and exquisite furniture which Indians generally do not have, which appeared to have been made by a highly civilized race. Also he described various strange tropical fruits these people grew, unknown to the outer world, on which they largely live. The colored man said he spoke the language of these people, since he spent most of his life among them; and he would bring me to them. It was clear from his description that these white Indians were really descendants of the ancient Atlanteans who colonized the highlands of Mato Grosso before the Flood, where they constructed their cities which enjoyed protection from the tidal waves that submerged Atlantis, whose coming they foresaw. Harold Wilkins treats of this subject in his two books on South America referred to above.

The man said these white Indians were very peaceful people, and were very intelligent. When I expressed my desire to visit them, his enthusiasm knew no limit, for it seemed that all his life he had waited for a foreign explorer to come to Guiba, as Fawcett had done a few decades ago, when his father had led him on his memorable journey to Roncador, never to return. While he lived, he hoped to follow in his father's footsteps and bring another stranger there. So strong was his enthusiasm that he said, "Tomorrow morning we will be off," and though a poor man, he scraped together his money and bought a gun for 3,000 cruzieros the same evening.

I visited the missionary again and told him about my contemplated journey. He was much upset and told me to bring the colored man to him. The black man claimed his father brought Fawcett to Roncador, while the missionary said that his three Indian guides confessed that they murdered him and returned. While in the sitting room of the missionary, the colored man kept saying to me, "I know a Great Secret, a Great Secret;" and then he pointed below, indicating that this secret pertained to the Underworld and certain openings in the earth that lead to it. Then he added, "This will be the greatest opportunity in your life. If you miss this opportunity it will never come again."

The missionary then entered and began talking to him. The black man spoke of his having spent most of his life among the white Indians of northern Mato Grosso and that he spoke their language fluently. The missionary also spoke the Chavantes language. Then ensued some conversation in Chavantes I did not understand. I presume it referred to Fawcett's fate. Then the missionary got furious. It seemed that this colored man's account of Fawcett's last journey did not correspond with his own claims, since the colored man said his father was Fawcett's guide, and mentioned nothing about Fawcett having been killed, while the missionary told the newspapers and everyone that three Indian guides confessed to him of having murdered Fawcett.

The missionary then turned upon the colored man and accused him of buying the gun in order to murder me in the jungle and rob me and said he would notify the police, so forcing the colored man to leave; and he warned me not to embark upon my contemplated journey for a month through the jungle, where my life would be in danger. So I was forced to abandon my contemplated expedition to Roncador; and, heartbroken, returned to Joinville. In the sequel I will tell how I discovered in the area of Joinville, in the state of Santa Catarina, Brazil, the great center of the tunnels and subterranean cities of the Atlanteans.

MY TUNNEL RESEARCHES IN BRAZIL

Early in 1957 visited Sao Lourenzo in Minas Gerais, the center of the Brazilian Theosophical Society which has an immense temple dedicated to "Agharta," the Subterranean World. I met the president of the Society, Prof. H.J. de Souza, who claimed to have just returned from a visit to the Subterranean World and to have possessed the "Keys of Shamballah."

He directed me to the Roncador Mountain chain in northeast Mato Grosso where Fawcett was heading when last seen on leaving Guiaba. I flew to Cuiaba, Mato Grosso, but could not go thru a month through a jungle infested by fierce Ghavantes Indians.

Little did I realize that the great center of the subterranean cities of the Atlanteans was in the more mountainous country of Santa Gatarina and not on the level plains of Mato Grosso except for the centers in Roncador, which are guarded by Chavantes who kill anyone who dares to enter without special permission.

The first great discovery was the subterranean city described in my letter to SEARCH last Oct. (1959 - this discovery is mentioned in greater detail in Raymond Bernard's article UNDERGROUND EXPLORATION IN BRAZIL, which follows this one...B.W.) This was confirmed over and over again, that subterranean men live inside that tunnel near Tujucas do Sul.

My investigator August just returned after 3 days with 2 subterranean men (who reappeared 5 times) and has taken photos which I will develop tomorrow and send you (referring to C.A. Marcoux - apparently these photos were never sent...B.W.). If they did not come out, because taken inside the tunnel by aid of the lights these two men carried, I will send him to take other photos. This is near Piedra Blanca. There is a metal door to one tunnel and an immense stone house capable of holding 500 persons. This seems GENUINE. There is a city inside this tunnel.

The third genuine tunnel is near Boc. in south Santa Gatarina, where my trustworthy associate Genesio just went and confirmed that there are men talking inside two of the three tunnels opening on top of a mountain. This is genuine. This makes three.

Next is the tunnel near the Paraguay border which has a door inside which opens and inside it is illuminated and bearded men were seen. My informant was disinterested and did not seem a job as a guide. This is genuine. This makes four genuine tunnels, and probably cities.

Now we come to the less certain.

First is the one near Curitibanos, where August saw the men and heard their music and was told that a tunnel exists, but hasn't seen it, though he saw the men. This seems certain. This makes FIVE certain places where subterranean people were seen.

Next is the tunnel near Sao Francisco do Sul where music is heard inside, and where men must exist to play the music. This makes SIX places.

August just returned from a trip to L, in Parana, where a subterranean man is worshipped as a saint. He comes up from a grotto, and other subterranean people come up. Last week hoodlums stoned them, so that the police are keeping guard and let nobody near. Restrictions may be off next week and we plan to take photos of these people and may befriend them and visit their subterranean city. This makes SEVEN PIACES.

There is also the long tunnel of P.H. where August was told that subterranean men were seen. But this case is uncertain and requires more investigation. It is at least certain as regards people living inside and we will investigate and report.

These eight places can be visited and I have special guides for each one, who know it.

I am awaiting the arrival of my friend with movie cameras and other cameras to photograph this great discovery. To date I just charted out the terrain but did not intensive research on any special location. I want Marcoux and Roy Smith here for that.

An old German book in old type, written by an early German settler in Santa Gatarina, who got information from Indians who preserved Atlantean traditions so THAT OF ALL PLACES ON EARTH, THE EASIEST CONNECTION BETWEEN THE SUBTERRANEAN WORLD AND THE SURFACE, WITH GREATEST NUMBER OF TUNN3LS THAT OPEN AND SUBTERRANEAN CITIES, IS SANTA CATARINA AND PARANA, BRAZIL.

UNDERGROUND EXPLORATION IN BRAZIL

For twenty-six years, two-and-a-half times as long as Ulysses, I have searched for the terrestrial gods I have felt within me must still exist - the Atlanteans, the Lemurians, and the Hyperboreans before them. I have felt that they have not completely abandoned humanity and dwell somewhere in seclusion. I thought I was looking for a terrestrial paradise where I could gather vegetarians and finer people and start a Utopia on earth,, but the coming of the Atomic age and the increase in radioactive fallout discouraged me.

THE BEST OF THE HOLLOW HASSLE

As fallout increased my enthusiasm about a terrestrial paradise in this poisoned world diminished and my attention was more and more directed to the mysterious tunnels that exist all over this part of South America, At the coast of Brazil they go under the ocean in the direction of Atlantis and connect this southern coast, from Rio de Janeiro down to Porto Allegre, with Ma to Grosso and continue on to the Andes. Who built these tunnels and why were they built?

Colonel Fawcett searched the high Mato Grosso plateau for lost Atlantean cities. But he did not consider the possibility of Atlantean nuclear warfare which poisoned the earth's atmosphere, and that they constructed cities not on top of the earth but under it, connected by tunnels, and with an air filtering and purifying system. The Atlantean cities which he sought on top of the earth in Mato Grosso were inside it. It is claimed by Professor Henrique de Souza, archaeologist and head of the Theosophical Society of Brazil that Fawcett entered a tunnel leading to an inhabited subterranean city where he is captive and still alive. For some time I have continued Fawcett's search for the lost cities of Atlantis.

About two months ago a party of explorers I sent out entered a tunnel in this same area (ie. Tujucas do Sul). It was completely lined with stone blocks. It was quite dirty and gave evidence of not having been used for a long, long time. After a day's hike they slept and got up to hike again. Now the tunnel was perfectly clean, giving them the impression there were inhabitants beyond. They walked a second day and slept. The third day of walking brought them to the sound of voices, speaking loud. This frightened them and they returned.

Some distance back a branch tunnel attracted their attention. Inside this they saw a little man who looked much like the traditional dwarf, with a long white beard. He didn't see them but again the group got frightened and returned to the surface and home. After telling me this I encouraged them to return and enter this tunnel. They did. Again it was a long, long walk but after two-and-a-half days they came to steps in the tunnel leading downward. At the end of the third day they came to an immense cavity with an illuminated sky. This gave off a yellow phosphorescence which illuminated below a city of houses. They saw many small men, women and children, who were crying loud enough to be heard.

One member of the party got frightened, so they returned. They are not anxious to return to the city of the dwarfs lest they be held prisoner.I believe the dwarfs do not want surface dwellers to publicize the existence of their city. This might lead to intrusion and trouble.

My researches have shown that in addition to subterranean cities not far below the earth's surface, there exists a Subterranean World of Agharta in the center of the earth. This subterranean world, whose capital is known as Shamballa, is well known to Tibetans and the people of Mongolia.

One of my explorers, an Inca, explored a 300-foot vertical tunnel, descending by rope. At the bottom he came to a door which automatically opened. Behind it stood an eight-foot Atlantean, protected by a plastic-like substance from radioactive outside air. He spoke to the Inca through a loud speaker, saying that he came from the center of the earth by means of an "electronic vehicle." The Inca was also told that flying saucers were sent from the interior of the earth to the outer atmosphere to halt nuclear tests. Charts prepared by underground Atlantean scientists indicate that the human race would not survive on earth for longer than ten years (if the nuclear tests continued at their present rate). All that can be done is save a few worthy individuals by bringing them into the tunnel opening we discovered, which is one of the four entrances to Agharta in the entire world.

Since It is my work to save a remnant of the American people, and since I will soon locate inside the transparent screen behind the secret door, (after a make-shift elevator is installed in the 300-foot vertical shaft leading to the tunnel) where I may breathe air free from radio-activity, if you wish to cooperate with me (here he is addressing Mead Layne.. .B.W.) and act as a U.S. representative and have faith that what I say is true, write me and state the extent to which you may help in this last minute effort to save a remnant of mankind from the universal destruction now in progress, to end in ten years.

CHAPTER SIXTEEN
An Interview With Timothy Green Beckley
About The Hollow Earth

Question: Thanks for granting this interview Mr. Beckley, I'm honored you've agreed. I wanted to talk to you about the Hollow Earth Theory. I consider your book *Subterranean Worlds Inside Earth* the best introduction to the subject, so maybe you can tell us how you became interested in the subject and what made you write the book?

Timothy Beckley: I have always been interested in the paranormal — the offbeat. The house that I grew up in was haunted by the spirit of a young baby. I know, because my mother and I heard the sound of a baby crying and followed little shoe prints in the snow to the back gate before they disappeared. into thin air. There were also poltergeist phenomenon — like lights going on and off and doors opening and closing. Once a plate fell to the floor without breaking, after sliding across the kitchen table.

Long John Nebel At age 10, I had my first of three UFO sightings. Two brightly lite discs circling overhead. One positioned itself across the street over an abandoned factory building before disappearing. It vanished as if someone had pulled a light switch.

I used to take books out of the library and bought a copy of Frank Edwards Stranger than Science when I was in the fifth grade I guess it was. I used to listen to talk show host Long John Nebel who was heard in 30 states over WOR in New York. I was to become a regular guest on the show years later. He was around a long time before Art Bell. He had all types of far out guests. He interviewed Ray Palmer and Richard Shaver via beeper phone (I have taped copies of this interview which I sell to Shaver fans).

I started writing when I was 13. I had a small publishing company that issued little pamphlets and a regular newsletter called The Searchlight, which was all about Shaver. You see Ray Palmer had the only newsstand magazine on UFOs. It was called *Flying Saucers From Outer Space* (later it was shortened to just Flying Saucers). Gray Barker of They Knew Too Much About Flying Saucers fame used to do a column for the magazine called Chasing the Flying Saucers. When he stopped doing the column I offered to do a column in its place. RAP would give me a free page of ads in exchange for doing On the Trail of the Flying Saucers which ran pretty much till the magazine went out of business. Of course, Palmer has been deceased a long time now but every once in a while I still speak to his son who is Ray Palmer — but not Jr. cause he has a different middle initial!

Question: So how did this put you in touch with Richard Shaver?

Timothy Beckley: Well of course for many years Shaver was part and parcel of almost everything Palmer did in the field. Shaver and I corresponded. I would get a letter from him every few days. Not very well written and on a typewriter that had not had its ribbon changed in years it seemed. I was 16 years old I think and Shaver would send me a box of his picture rocks like clockwork every week or so. My mother dreaded the arrival of the box as the rocks were right out of Shaver's yard. The box was filled of dirt and worms and god knows what.

Shaver, Palmer and I had a lively debate going for a while over the reality of the subsurface world. I put a lot of this material together into a book that was published by Barker's long since defunct Saucerian Press. *The Shaver Mystery and The Inner Earth* went through several printings. By the time I was 17 my career as a writer was itched in cement I guess you could say...even though I had just about failed English in high school. Today our companies Inner Light Global Communications have about 85 titles in print by authors like Commander X, Brad Steiger, Tim Swartz, T. Lobsang Rampa. A bootleg edition of The Shaver Mystery is sold on the net by Health Research. Have never seen a cent on it so if you want the full story get my book *Subterranean Worlds Inside Earth* which has all this original material plus a lot more. It is available from Amazon or from our own website www.conspiracyjournal.com

THE BEST OF THE HOLLOW HASSLE

Question: Most "UFO researchers" today might consider the Hollow Earth theory a "crack-pot" subject, but since the beginnings of the modern UFO sightings it has been intimately connected with "UFOlogy" — especially early contactees. Why do you think this has been forgotten?

Timothy Beckley: Unlike most folks who consider themselves UFO researchers and paranormal investigators, I have never had a deep set belief pattern. Some UFOs may be interplanetary craft, but the majority has to be placed in some other category. During radio and TV interviews the host will always ask what is MY opinion about UFOs. I always tell him it doesn't matter what my opinion is — as UFOs act independently of my beliefs or anyone else's!

What really got me thinking about this inner earth stuff was back years ago there were three miners trapped in a cave in Pennsylvania. It was all over the newspapers and radio back then. They were not expected to survive after three or four days of not being heard from. They were eventually rescued. In the hospital told about after two days or so they saw an eerie light in the cavern where they were trapped. They followed it and came across a door that lead to beings who gave them food and water so they were able to stay alive.

There are so many legends that tell about miners trapped in caves who have experiences with a race of short beings.

To this day I don't know if Shaver was simply repeating these stories that he could have heard, or if was really in touch with these tero and dero.

Shaver's writings were titillating and well written. Palmer probably jazzed them up as I never found Shaver to be a very good writer. Palmer was a real pro. Some say he was responsible for starting the flying saucer mystery with the Maury Island "hoax." If anything he did keep the mystery alive during its lean years. We need a few more Palmer's in the field today. I'm afraid today's breed of self professed "experts" are a wee bit boring!

I'm afraid I am a bit of a fence setter. All this could be real or none of it could be! But we should investigate such matters. Beats watching TV 18 hours a day.

Actually, I am a bit more predisposed to believe in the Shaver Mystery rather than accept the idea of openings at the North and South Poles. Not totally convinced of this theory yet. Want to see more proof.

And while Shaver and Palmer are long gone since gone from our realm their spirits definitely live on. There is still a great deal of interest in what they wrote, and a whole new generation that wants to be exposed to what they had to offer the world. If I have my way I will continue to be a source for what they had to say!

Question: You related some good miners' stories in your book as well. This is one of the most convincing parts of it in my mind. Someone should go out and interview a bunch of them. Has anyone compiled a book strictly on the tales of miners in the tunnels?

Timothy Beckley: As far as I know there has not been an attempt to compile all the stories that miners have told about strange things happening underground. But in my various writings I do quote from some of the sources that are known to me.

The well known conspiracy author Branton (who once wrote an index to the inner earth under his real name Bruce Walton for Saucerian Press) has quite a bit of this information on his various web sites.

By the way, I just got word that Branton was in a very serious accident. Apparently he was riding his bicycle when a truck plowed in to him. He is, at last word, in intensive care — but I don't have the information as to what hospital he might be in.

We could use a few active researchers to explore the old journals and books for more exploits by miners and cave explorers. There is probably a ton of uncovered material out there just waiting to be discovered. Unfortunately, we don't have someone with the influence like Ray Palmer — or his magazines — to attract attention or "get the word out." I bet there are all kinds of small circulation publications put out by amateur cave explorers that would provide us with lots of new stories.

Question: In India popular belief has it that tunnel entrances are guarded by elementals. Kirk of Aberfoyle in the 1600's wrote of the cave dwellers with similar occult overtones. What's your take on these theories?

Timothy Beckley: Almost every country has its stories. India is no different. Many, many years ago Dr. Robert Dickhoff used to write about some of these matters. He did the book Homecoming of the Martians which gave very sinister overtones to the UFO mystery. In many legends the entrances to the caves were guarded by the Genies — who somehow later ended up in a bottle that you could rub three times and have your wish come true.

Question: I know you've managed to take a photograph of one of the Men in Black. This is said to be the only one in existence. Maybe you could tell us how that came about and what if any is the relation of the MIBs to Inner Earth Traditions?

Timothy Beckley: I was with Jim Moseley when we took the photo of this MIB. He was standing in a doorway in Jersey City dressed all in black just like Albert K. Bender had described in Barker's They Knew Too Much book.

This MIB was standing directly across the street from the apartment building where Jack Robinson lived. Jack was a freelance editor on the staff of Jim Moseley's Saucer News.

Jack had collected a lot of Shaver material. In fact in my book *Subterranean Worlds Inside Earth* he tells the story of a Steve Brodie who lived in the same apartment complex as Jack and his wife Mary (NOT the one where the MIB was sighted standing near). Brodie was a hermit of sorts who had no friends. Jack ran into him in the hallway one day and Steve invited him into his apartment where he proceeded to tell him of an experience out west that involved being in a cave and seeing creatures like what Shaver had claimed existed. Brodie was last seen on a train in Arizona looking like he was in a trance....and as if he were returning to where he had this encounter with the dero.

Jack told the story numerous times on the Long John Nebel show and wrote quite a bit on the Inner Earth for my newsletter *The Searchlight*...Bender always talked about UFO bases at the South Pole. So there is a definite connection here. I can't say for sure who this character was Jim and I photographed. But the picture is published in my book The UFO Silencers: Mystery of the Men in Black.

Question: Some of the amazing discoveries of Tesla and John Keely bear a strong similarity to Shaver's descriptions of the Deros "mech-ray" technology. Did you ever talk to Shaver about these inventors?

Timothy Beckley: Unfortunately, in the last days of his life Shaver didn't seem to want to talk about the Dero and the Tero or their Mech Rays Technology — and I was not that knowledgeable about Tesla in those days (remember I was only 16 or 17 years old when I corresponded with Shaver). By that time he was pretty much into his rock phase. I was interested in the rocks but you have to have a bit of patience to work with them. By the way I don't recall if I mentioned it or not, but I purchased the last 50 copies of The Secret World that Palmer and Shaver did together on the rock phenomenon. Its available either directly from our website or you can order it off of Amazon. Toward the end, Shaver had a very bitter attitude toward most people as he felt he has been used to a large extent, in that people only heard what they wanted to hear and believe what they wanted to believe. Shaver believed in a physical world beneath our feet. No astral projections, no ghosts, no demons. Pure and simple Dero and Tero without the frills.

Question: Of the UFO researchers today, who do think is on the right track? Does anyone carry on the Shaver tradition?

Timothy Beckley: Well I am not sure what the "right track" is — I think of myself as a journalist — a reporter. You have to retain an open mind while at the same time be critical when an account or a story just doesn't add up.

I am not really in touch with other researchers outside of Branton, Commander X and Tim Swartz. Branton was, of course, seriously injured in an accident a while ago and I don't know when he will be making a comeback. His books Omega Files - Underground Nazi UFO Bases and Dulce Wars seem of interest to the public. And Underground Alien Bases by Commander X gets an increasing number of hits on Amazon.com.

Who else is really writing about subject — and not rehashing material? I would like to do an update on the Shaver mystery and the inner earth. A lot of my material was ruined in a flood a couple of years ago. I would like to get some of the better letters and stories from early Shaver and Palmer magazines if anyone has anything to lend, donate or email. First hand or passed down accounts also would be welcome. Anyone reading this can always email me at MRUFO@hotmail.com, and maybe we can put together something in the next year or two. Anyone contributing material will receive a free copy and credit.

Question: That's a great idea! I would love to see some more Shaver stories. Any last thoughts? What's in the future for UFOs?

Timothy Beckley: I wouldn't think the future of UFOs is either bright or bleak....it just is! I know that may sound pretty stupid. But, UFOs do whatever they do without much interaction on our part — unless they want interaction. I am afraid it is pretty much a one way street with them coming and going whenever they want to come and go. At times they pay us little attention and I suspect they could not care too much if we know they are around or not. There are UFO reports all the time, and monster sightings (do not forget to see Mothman Prophecies when it hits the movie houses next year, based on John Keel's book) and tons of weirdness to keep any modern day Charles Fort busy 24/7.

I doubt if the field will ever obtain the level of excitement that it did in the past. But that is maybe just because I am not easily excited — except perhaps by a hot blonde from Talos — these days. It seems those that are getting attracted to these fields retain the same degree of fascination I might have had when I read my first article by Palmer or Shaver and purchased FATE for the first time. I keep at it. I will not rest. Its part of my life.

I hope everyone enjoyed what I had to say and good luck all of you in your work to reveal the truth, whatever that might be.

To find out more about Mr. UFO, Timothy Green Beckely, and other hollow Earth mysteries, visit his website at:

www.conspiracyjournal.com

CHAPTER SEVENTEEN
No Smoke Without Fire
-OR-
The Last Of The Sun Worshippers
By POOSHKA
Vol. 5, NO. 2, Feb. 1984

As In love, I had rather love too many than too few, so in charity, I had rather Believe too much than too Littie.
- St. Gregory

Given the climate of disbelief concerning Hollow Earth (HE for short), let us begin instead with the small mountain of evidence, now forming, which addresses itself to the mysteries of the inner earth. To do this, one can start anywhere, but most start at the North Pole: Many fascinating stories of men, ships, tribes, who broke through the ice barrier into a lush and enchanted country. Compelling as these accounts are, they have largely drifted down to oblivion - tossed aside by the unfavorable intellectual climate - or is it a snake in the grass?

Without the stamp of authority on these accounts, we are left with circumstantial evidence. And so, circumstantial evidence it will be! And what a legion it is! From the standpoint of Science, there are the results from the space (satellite) program; Soviet and other eye-opening studies of the earth's conformation; not to mention the exact science of the earth's interior, the basic laws of nature that prevail within, the question of matter and energy as it permeates the inner world.

A great heresy Hollow Earth is; even UFO and Monster buffs shy away. All HE channels have been over shadowed by this. In circles where conspiracy theories are popular, HE can l'incl a niche but it is wise not to linger too long in this sanctuary of the Age of Disbelief. How we worship theories - and dread realities!

How bold to speak of realities, but let us not be drawn away by conspiracy theories -or semantics. Fiction and faction are beginning to sound alike. In a surprisingly short span of time, we started taking Science Fiction seriously, wondering where such ideas came from. Many a fine researcher masquerades as a paperback writer. Let us say that food for thought - or icing on the cake - comes from such pleasurable reading experiences. And let us give the discerning eye enough credit to sift greater from lesser realities. We are at this point in HE; an army of evidence stands alert, ready for resolution into the great ocean of truth. Readiness is the solvent. But who will take the plunge? Thousands of years of history have conspired to draw the veil over the hidden world. What invalidation, ridicule and outright suppression will not stamp out, the mind of man holds no court for Inner earth.

How much do we really know about this earthly bubble that floats in cosmic space? The densest portion, the crust, has two natural explanations. Untold eons of spin have thrown out the heaviest matter toward the periphery (we might imagine, then, that the finest, most subtle matter remains at the core). Considering also the accumulation of cosmic dust, the debris settled upon our surface through the march of time/space, the earth has thickened its "skin." The theory of continental drift falls naturally in relation to that of an expanding planet, simply busting out of its britches. It seems quite likely that a planet, like other living things, grows. Certainly Geology, Archeology, Paleontology, etc. know all this. The stuff underneath is old, underneath it, older; the further you go the greater the antiquity.

Why abandon these basic principles at "bedrock?" Science knows that at approximately seven miles above the surface of earth the troposphere ends and the stratosphere begins. But Science has not been the same seven miles down. Advanced as our technology seems, we have not penetrated one-thousandth of the earth's depth! (But look out for the Russians at Azerbaijan!) Beyond this point, our information comes from other channels: miners, aviators, navigators; speculations from seismology; myths of autochthonous origins; accounts of travelers, explorers, wayfarers; contacts made by adepts and clear channels; etc.

As this almanac of inner earth grows, one simple fact emerges - the great antiquity of our hollow earth. Delving it, we start with an antediluvian world and work backwards in time. Some say that Destiny is prepared to visit our planet, breaking the seals of the book of time. If so, it might behoove us to make preparations. There is a Muslim saying, "Trust in Allah, but tether your camel first."

There are many ways to tether hollow earth. Returning to literature for a moment, let us look at some of the men who wrote the books. Sir Rider Haggard was the author of many popular romances (KING SOLOMON'S MINES, SHE, etc.) which more than hint at the existence of an inner earth colony. Both Rudyard Kipling and Madame Blavatsky believed that SHE came "through" Haggard, rather than simply written by him. These British gentlemen, including Col. Percy Fawcett, were, in any case, on the same wavelength. The man Fawcett may not be familiar to this generation. Fawcett, the explorer, soldier, engineer, artist, archeologist, and cricket champion! A contemporary of Haggard's at the approach of the twentieth century, Fawcett and his Brazilian expedition are still being studied 60 years after his disappearance in the Roncador mountain region. The Colonel and his two companions were hot on the trail of a lost city, fondly dubbed "Z," when last heard from. The public outcome was inconclusive, despite strenuous efforts to make Fawcett look dead. Indications from his inner circle, however, looked more favorable.

Before departing for Brazil to embark on his final expedition (1925) Fawcett was given a small artifact by the novelist Haggard. It was a 10-inch-tall black basalt figurine with inscriptions on its chest and ankles. The object had a peculiar electric current; some people weren't even able to touch it.

In the hands of a sensitive, the figure recalled the effigy of a high-ranking priest, officiating in an elaborate temple hewn from the face of a cliff. The leader of the priests is wearing a breastplate similar to the one on the figure. In this vision, the high priest is handing the relic to another priest with due instructions to retain it carefully, until it is to be passed on to the appointed one. At length, it comes into the possession of one who embodies the original personage "when numerous forgotten things will through its influence be elucidated." (*Lost Trails, Lost Cities*, by P.H. Fawcett), Many details of the Fawcett adventure suggest that he was that man. Whether our psychic was reading these impressions from the black figurine - or from Fawcett's own thought waves - matters not. For the 18th century record of a Spanish expedition into "Z" mentions that in the center of the main plaza stood a larger than life figure carved in black stone, matching the features of the 10-inch figurine.

Arthur Conan Doyle, who also dabbed in these things, attended a lecture given by Fawcett in London, in 1911, and from it took inspiration for THE LOST WORLD, his epic of an antediluvian remnant sunken in the deep impenetrable forests of Brazil.

Lost trails, lost cities, lost worlds. Those who have been on the inside say it is easier to get there than to find your way back. So, if our camel has strayed into the cavern, without returning, we have not a good enough tether.

Perhaps it is an obscure form of logic to try to tether HE with an equal mystery. But an unbroken stream of disappearances demands this. Not to mention the Bermuda Triangle or the perfectly arranged vanishing points worked out by Ivan T. Sanderson and his team of experts, the question of vanishments leaves an untidy fringe on the pages of history. Sure, there are plenty of ways to go. But we are concerned here only with those cases which left a trail - to the subterranean world.

Both ordinary and remarkable people have vanished, stirring the imagination to wonder about "certain histories of children and men and women who vanished strangely from the earth. They would be seen by a peasant in the fields walking towards some green and rounded hillock, and seen no more on the earth."

An Indian story from Oklahoma, where the mysterious mound-builders once flourished, says that an odd scroll was found near one mound, concerning a 16th century Spaniard who had been part of the Coronado expedition. This man became separated from his company and somehow came upon a subterranean entrance, into which he ventured, never to return.

Disappearances connected with caverns and HE research are too numerous to mention. Certain areas, like California, seam most susceptible. Near Death Valley, for example, an Indian trapper tells of his grandfather who lived in a "strange country" for three years, beneath the Panamint Mountains after having entered a tunnel in Emigrant Canyon. In Northern California, around the Sonoma area, there have been several unexplained disappearances along a stretch of road 110 miles north of San Francisco in lake country of Mendocino County. "Even trucks have vanished...all the U.S. Government's. The U.S. Government has noted the areas as rough, unsurveyable and UNEXPLORED." Notorious for anomalies, this region is among the leading "entrance points" in the U.S. In the thick of it - a region which has no wind, where forests fires won't burn, where the silence is so perfect "you can hear insects running on the ground" - in its midst is a cave that "has steps leading down and there is no sound when a rock is thrown in." The man reporting all this was a cattleman homesteading on government land; but "too many incidents to be told...All in all we stayed there about two years before we quit." (from Bruce Walton, INNER EARTH ENTRANCE SERIES)

The cattleman, evidently, was not quite ready to disappear. Another focal point or vanishing point has long been remembered in European folklore. The Celtic lands are rich in such lore concerning the disappearance of ordinary folk. The Pied Piper of Hamelin, immortalized by the poet Robert Browning, is the classic tale of 130 children who disappeared into the Koppelberg Mountain (hill), led by a stranger, a demon-piper, a man who came out of nowhere, and, in retaliation to a broken promise, abducted the children of Hamelin into his home inside the hill. A curious footnote to this tale is recounted in the last lines of the epic poem, written in 1888:

> And I must not omit to say
> That in Transylvania there's a tribe
> Of alien people who ascribe
> The outlandish ways and dress
> Of which their neighbors lay such stress,
> To their fathers and mothers having risen
> Out of some subterranean prison
> Into which they were trepanned
> Long time ago in a mighty band
> Out of Hamelin town in Brunswick land,
> But how or why, they don't understand.

Not until the present century, though, have such disappearances been duly recorded for our scrutiny. Kentucky (Mammoth Cave area) has long been a favorite for hollow earthists, as it holds the entrance to ETIDORHPA, the classic American journey into the earth by the Man who did it. Reaching a distance of some 700 miles into the earth, the journey ends, almost abruptly at this, at The Unknown Country, The End of Earth.

(I have a friend, who in the course of his researches into the Tunguska Explosion (in Siberia) made a careful study of news reported in the year of that blast, 1908. This was an interesting time

"There was a Nixon, an Agnew, a Castro; you don't even have to look for it. Nikola Tesla was at work at this time in Colorado Springs. Also in June, 1908, a lot of people disappeared in Kentucky...")

Twenty years later, another legion vanishes, this time back in the Roncador Mountains, just three years after Col. Fawcett's disappearance there. Carl Huni: "When I was in Brazil I heard a lot about the underground caverns and subterranean cities...A good part of the immigrants who helped in the uprising of General Isidro Lopez back in 1928 disappeared into these mountains and were never seen again...Finally they made a truce and let the 4000 troops go; about 3,000 of them went to Acre in the northwest and about 1,000 disappeared in the caverns. I heard the story consistently."

The above is quoted from THE HOLLOW EARTH by Dr. Raymond Bernard, A.B., M.A., Ph.D., who himself disappeared in Brazil in the late sixties. Notwithstanding savages and beasts, there is a mysterious district along the upper reaches of the Amazon River in which a number of explorers, from the U.S. and other countries, have gone and not returned.

This opens up the question of the disappearance of remarkable men:

-Fawcett and Bernard, inner earth explorers of this century.

-Count St. Germain, into the Himalayan mass, at the close of the 18th century, (he promised to return)

-Ambrose Bierce, the satirist, disappearing under similar circumstances after becoming interested in the Crystal Skull.

-Doc Anderson, "the man who sees tomorrow," a known psychic who was initiated into the Asian pyramids; and promised to use his powers to locate tunnel entrances in the U.S.

-William Beebe, the oceanographer who expressed his belief in the coming invasion of the earth by an underground race.

Of remarkable men who have gone, and returned, there is the account of the Iroquois Chief, who entered the Endless Caverns of Luray during the Civil War, to insure the safety of himself and his tribe. Entering a small opening near Sweetwater, Tennessee ("the Lost Sea of Tennessee still carries great unexplained mysteries"), the group did not surface until nine years later. They described their stay at God's Teepee that existed below a great sea deep within the earth.

These are not random cases. History demurs on cases of whole cultures slipping through her back door; but has no answers. Entry inside the earth is highly suspect considering native lore and the trail of evidence left by the ancients of Tibet and Afghanistan; the abandonment of Angkor Wat in Cambodia; and the quiet departure of the Egyptians. In more historical time, there was the patent disappearance of thousands, probably millions of the races of Inca, Maya, and (Colorado) Cliff Dwellers, each with extensive tribal knowledge of access to subterranean habitations, each making a clean break with the enturbulated surface world. Finally, there are the Lost Tribes of Israel, who, according to a now sizable literature, may be placed in the enchanted country inside the North Pole.

Enough of our negative argument of disappearances; promises to return are often tagged on to these stories. Let us wait and see.

As for positive arguments, let us move into an even graver heresy, but one which squarely confronts the proposal of hollow earth. Based on experience inside the merest crust of earth, it is unanimously believed that the inside of our planet is utterly dark, a rayless doom. Down under, there are, of course, the blind, white fish of the caverns. The Bathynelles, who have the longest pedigree of all living organisms. And Proteus Anguinus, known also as Little Dragon, who eye consists only of a crystalline lens, who has no tongue (contriving to use its thyroid gland as a substitute) and who, though blind, can detect light and flee from it. It accomplishes this through its skin, through a process of negative phototropism which is little understood.

But, it is argued, without light there is no life. The words are almost coterminous-light is life itself. And beyond the curious troglodytes of the caves, there is little subsurface life known to man. The matter rests there.

To contest this brings us to the ultimate heresy - a light within. The Central Sun. Light within, as proposed by our spiritual guides, is no mere metaphor. We earthlings are but a small model of the Great Spirit, our Earth. O' self-deceiving mortals of the surface realm, disbelieving the inescapable, believing vanities, worshipping gods that have failed you!

There could be no hollow earth without light. And the HE cabal is ready to take on this issue.

The Talmud says that Noah had no other light in the Ark than that furnished by precious stones. (Was the Ark, after all, underground?) Stones such as the carbuncle have wondrous powers, able to fill a whole room with light.

It is said that the King of Pegu had a carbuncle so bright that in a dark place it made all bodies transparent. The Caliph of Babylon is said to have found in the pyramid of Cheops a carbuncle the size of an egg shining like the light of day a recent under-earth episode describes the journey of one John Deitrich into a place with wonderfully brilliant walls - it was quite clear and bright, and yet there was no heat; "the large carbuncle set in the walls and roof gave light instead of sun, moon and stars."

Earthlight; who can resist the temptation to pry into its mystery? It is unusual to read any lengthy account of the ongoer's journey through the earth without finding mention of this "strange light." Lucian (in VERA HISTORIA) describes it as the "soft twilight of the dawn in an eternal Spring" in his account of the underground rites of Tammuz. Subterranean light is most often described as a soft green luminescence; sometimes seen as dancing, sometimes phosphorescent, sometimes ubiquitous, perpetual - no night, no day, no evident source of emanation.

Although the green glow has been observed as far down as seven miles into the earth, at times it appears at a much greater proximity to the surface, as at the edge of the Sierra Pacaraima, in Venezuela. At night the volcanic crater here gives off a soft green light that illuminates the trees at the rim of the volcano. The Kingdom of the Two Craters, - for there is a pair of volcano's with an underground tunnel connecting them - was examined in the sixties by two noted archeologists. Of course, the Venezuelan government clamped down on information. However, there were additional rumors of a strange people inhabiting the volcano. Ironically, the nearest half-breed town to the twoin craters is named "Esmeralda," (Emerald). For two nights before the expedition moved in, there were signs of an evacuation.

This is not too strange to hollow earthists. Inhabitants are usually part of green light stories. The "green country" under English soil is said to be entirely green - the people, the animals, the earth, and sky. "There is nothing that is not green. The sun never shines...The light there is a constant green glow, as if the sun were always just below the horizon."

The "soft green light" of Venezuela, mentioned above, is sometimes called "money light." Here its proximity to oil regions, suggest underground gases. It is interesting that the glow is dubbed "money light", not because money is green, but because it has long been regarded by prospectors as a sign of the nearness of gold. What causes this association is not known.

The green light has been seen in Azerbaijan, a green fluorescence emanating from a "bottomless well." Curious that the Russians are deep-drilling in this very place which was once Persian land.

Finally, the learned lamas of Tibet and Mongolia spill the secret of the ages in opening the book of Agharta, the cavern world within the earth, home to millions, with extensions even in the great American caverns. The green luminescence of this abode sustains plant, animal and human life, yea nurtures them to a dimension of growth and longevity inconceivable to surface man.

The permeability of earthlight is remarkable, for in certain places the walls themselves appear to be the source. There is no heat with this light; nor is there an apparent fuel source - rather its rays seem to come from everywhere.

Luminescent plankton, found in the seas of the Indian Ocean may be stimulated to light up by sonic waves at a frequency out of human earshot. An underground complex near Death Valley also attributes illumination to sound waves; the confines of Yaktayvia, according to this report, are illuminated by a system of "chimes." It sounds like we are talking of artificial light, then.

One theorist suggests that the "eerie" lights of Matto Grosso (Brazil), coming from crystal "light pillars", bear the mark of a people who knew how to "eternalize" a cold form of light." Tracing these sun-worshippers through the Phoenicians, the Basque; through the Glostershire uplands of England and the Furetan pillars of the Colombian highlands, as well as through Maya and Egypt, the theory implies a once-universal knowledge of this form of light.

Recently a similar artifact has come into HE news with the exploration of Blowing Cave near Cushman, Arkansas. Charles Marcoux, who at this moment may be within the corridors, claimed there is an ancient relic, discernible near the entrance of Blowing Cave, a beacon light (which he caught on film). Seven miles down, Marcoux has a date with destiny, (see "Straight Talk" in past issues of The Hollow Hassle).

If we are on to something, these light pillars of the Nile, the Terrible Crystals of the Scriptures, were the very means whereby the Egyptians accomplished their artwork within the darkest chambers of the pyramids. Though some of these chambers are generously carpeted in guano, the inner chambers showed no soot upon the walls. What manner of "clean light" was used by the priests-artists?

The Egyptians had a few secrets. Cryptologists have opened Egyptian tombs sealed for thousands of years which contained glowing lights, generally described as "globes on pedestals", emanating no heat and "from no observable power sourcell. In Italy another such tomb was opened containing the perfectly preserved body of a thousand year corpse, a very pretty young woman who seemed to pass from life only moments before. Similar tomb lights have been found in Greece and Syria.

We are the last of the sun-worshippers. Old gods die hard; fugitives of inner earth can only await a new science of man. But hollow earth is calling and there is still time to heed that call.
Romancing about extra-terrestrials, we dare not even give a thought to "inner-terrestrials," our subterranean neighbors. There is glory in the stars, but danger down under. We are afraid to know our Earth, to enter into the Great Spirit. But in time, the myths will die, the "fiery furnace" will be extinguished; and the light will shine, in its fullness, its penetrating splendor; its gentle

approach will remove all doubt, and the world will, once again, receive an old friend.

CHAPTER EIGHTEEN
Crossroads: Letters To The Editor

CROSSROADS

From Vol. 3, NO. 1, Oct. 1981

Dear Mary,

What a trippie issue The Hollow Hassle was!! Usually they all are, but for some reason this last one hit all the items that are passing through this area. Floria Benton's "Polar Shifts" has brought another memory to life. This was on the KNPT Talk Show. The guest speaker was a man called Ron (can't recall last name). He has written a book called "There's No Place To Hide" and he believes, as do many other psychics, that around the year 2,000 A.D. something big will cause a drastic change in the earth. Ron has a friend who's a scientist, who predicted all the facts on Mt. St. Helen, but most people just ridiculed him at the time.

Ron wrote his book on the information provided by this scientist and it goes like this: "In Douglas County, OR. there's a sleeping mountain called Crater Lake. At the widest point it is five miles across. This volcano is cylinder in shape. Most are cone shaped (the throat). There's a granite and basalt plug blocking the throat of Crater Lake. This plug is one mile wide and consists of pillars (30,000 of them). They range in size from two inches to large enough to park a full rigged Mac Truck on. Over the last several hundred years this mountain has been going through stages of maturing, and according to the scientist, around the year 2,000 A.D., will go into its final stages. The pressure from below and the water above will equal and then the plug will start to lift. As it lifts, the water will slide in around and go down the throat.

Meeting with all that heat (500 plus degrees), it will change to steam and create a pressure so great that the plug will come out like a rocket missile and as it does, it will come apart. Pieces will range from powder to pieces as large as a two-story house. Giving a mild estimate of what there could be in number of pieces puts it at 1,000,000 pieces (as big as a house). Some of these pieces will go into orbit and down on the earth - who knows where. The explosion from this will be enough to knock the world a bit cockeyed."

Another man from Newport, by the name of Virgil Godwin, was also on a local Talk Show. He is a UFO buff, also has interests in Bigfoot. From what I can gather, this fellow has written a book on UFO's and is also a photographer. He has invented a special camera and develops his own film footage and pictures. He says he has thousands of pictures of UFO's, has been inside of them and met the people who man the ships. He claims that the USA is one of the major bases for this underground secret order and that these people are not from space, but rather from here on (or in) Earth. They have tunnels!

He says that they are just like we are - only the women rule. They are eight-feet tall - the men are much smaller. Their ties and background are American Indian! He also said that there is much he can tell about them, as they trust him, and he knows where they are located. Lincoln County is supposed to have quite a nest of them. They are very skilled and Mr. Godwin has even seen a top one-half of a mountain slide over, so ships can land or take off.

As for Bigfoot, they are a domesticated pet to these people - they axe watch dogs. He has seen the area in which they are kept. The people speak English, but their tongue is some dialect of American Indian. In passing, Mr. Godwin said he felt sorry for any country that attacked and bombed the U.S., as we axe their base and they will retaliate. One would also be wise to check out any areas where Bigfoot have been spotted. If they are watch dogs, they must be watch-dogging something!

Shu - Waldport, OR.

Ed: I want to thank Shu for her very informative letter.

THE BEST OF THE HOLLOW HASSLE

Dear Mary,

Please find enclosed a check for a year's subscription to The Hollow Hassle. I am hoping it will prove useful in my own studies. I have been working on a biography of Cyrus R. Teed and have read the majority of his works. Because of it I have gained an interest in putting together a collection of other hollow earth theories. I am hoping that you and your magazine might be a knowledgeable source of useful information - leading me to other "hollow" thinkers and their books, histories, and ideas. Any information would be greatly welcomed.

Bruce Pennington - Brockport, N.Y.

Ed: Welcome aboard Bruce! We'll be glad to help you in your research, in any way we can.

Dear Mary,

New Mexico and Arizona have the most complex of united cavern systems in the USA, which go into Mexico too. The cavern inhabitants are not all beneficial, but most are not destructive to surface people, and they do not interfere with us to any great extent. Outside of deros and rays, the main thing is that you may get a contact, if and when you are near them. You will either think it is a spirit, or else you'll think you're crazy, while they are only trying to let you that they do exist in the caverns. Here are some places that could be searched:

Go about 14 to 20 miles northeast of Pie Town, N.M. Go east through Pie Town on the paved highway, about 14 to 20 miles. Somewhere along there is a gravel road that turns north to "El Moro" National Monument. As you go north on the gravel, you'll see a fence on the east side of the road. When the fence comes to an end, park and walk along the fence to the east. Look for a FLAT SQUARE ROCK on your left, as you walk east (watch for snakes). The ground has caved in under one corner of the rock, and you can squeeze into this hole. A good light is needed, and you will find yourself walking down man-made wide steps that go down into a large hall. In the center is some kind of altar of a round shape. Further on are more steps going down deeper into this cave. While hunting in this area, a man found this cave, but they didn't have any lights, only matches. He planned to come back later, but didn't return until a few years later. He was so curious that he couldn't get it out of his mind. To make it short, he couldn't find it again, no matter how much he searched - it was futile.

THE BEST OF THE HOLLOW HASSLE

Now, around Farmington, N.M., and also along route 666, in the mountains on both sides, I get strange sensations (ESP) that need to be explored. The Organ Mts. in Las Cruces, N.M. and the Franklin Mts. in El Paso are also connected with the cavern world. There are the Black Range Mts. and the old ghost town of Mogollon, N.M., which I have explored and photographed the unseen worlds around this area.

There are "white rays" about 33 miles southeast of Mexico City, but don't count on getting in, unless they see your mind and attitudes are of a high consciousness. There is a "fountain of youth" within 50 miles from Deming, N.M. and it goes into Mexico. The Aztec, Toltec, and the "degenerate one" are at war in the caverns. Much of this activity extends into Arizona and N.M. against the red and black Indian legions in the caverns, and war is also causing much of the "ray tamper" on the surface. I am also aware that the Bermuda Triangle conditions are involved with the present problems in Florida, and Florida will, eventually, sink below the sea because of it.

There are hundreds of places to investigate. Many cannot be entered, but some are safe and can be gotten in to, providing that your mind shows no ulterior motive. As a closing, I advise all explorers to be cautious, be prepared, and be careful of who you are taking with you, for you can find one of the greatest experiences of your life. Once you have felt the "rays," you'll be stimulated and elated in such a way that you cannot imagine. It is, in a small way, like being entranced, for being near or experiencing a white ray will cause you to be "EN-TRANCED." Are you ready for that experience?

Charles Marcoux - Phoenix, AZ.

Ed: I can't speak for anyone else Charles, BUT I'M READY!

Dear Mary,

Did I tell you that I have reason to suspect that a lost city in Greenland is an abandoned Nazi base, secretly built there during World War II? I first read about this back in the fifties. It was in a 1954 issue of FIX, which used to be a feature zine. That issue also had photos of a strange tribe in the jungles of Borneo that had hairless 10 inch long tails. It came out of where their backbone ended. They were a medical anomaly. Anyhow, I will fill you in on any new info I can find on the above Greenland city.

Bob Schiller - Denmark, Wis.

THE BEST OF THE HOLLOW HASSLE

Ed: Bob, I was quite interested in your Greenland info, as some years ago a friend of mine was forced to fly over the southeast part of Greenland, due to bad weather conditions in the North Atlantic. He and the other passengers on the plane were amazed to see, in a valley below, large pyramids. My friend has flown over the pyramids of Giza and claims that the pyramids in Greenland were much larger. I never was able to find out anymore on this though.

From Vol. 4, NO. 1, Oct. 1982

Dear Mary,

I like your name, Hollow Hassle. I'd like to know more about your paper, magazine or whatever. The grapevine didn't say much about your setup, just your name and place. By the time you get this letter, I'll be up north Canadian area, but I'll be home in a couple of weeks. Once I leave home, I can't be reached, but I will answer if you write.

If I'm right, you're interested in what amounts to, "A world within a world." I'm interested at the present time with the "Horned Rat Mole." I think this is the largest form of life on Earth, including the sea life. These moles are extremely dangerous and cunning. Maybe you have information on them. If you do, then you have heard of the Chinese Eleven (C-ll) group, the Russian headache, as over 500 specially trained Russians spend all of their tine hunting the C-ll group. I trained the Chinese group years back, to keep the Russian government screwed up. At this time, Mary, our government has a secret law. They claim ownership of all the information pertaining to the Unknown. They have us from both ends.

If we tell them anything, they ignore us or degrade us, or commit us to a nut factory. But, we have a group in Washington, D.C. that is seriously concerned with the Russian 500 group. Whoever solves the mystery first can rule the world. This is no joke! I have no way to educate our people, but one person can't do the job alone. If you have time Mary, drop me a line.

Virgil T. Godwin - Salem, Oregon

Ed: I definitely did drop Virgil a line and am awaiting his reply. I have never heard of the Horned Rat Mole or the groups he mentions, but if any of our readers our familiar with any of this, please drop me a line.

Dear Mary,

I'm sending $2 for a sample copy of your quarterly issue. I've been involved with a group interested in going there, which somewhat fell apart. I've gone to D.C. and looked through the polar archives. When I asked for Adm. Byrd's diary, they said it was still in the hands of the family, until the settlement of the estate. Well he died 20 years ago.

Anyway, our group decided the best way to go there was a boat to the North Pole. There do seem to be some caves in South America that are entranceways. Also, I've read that account called "Etidorhpa," about a man who took a Kentucky entrance and was lead quite far in. Also, did you know about the bill before Congress, (HR 3973), which would - 1. Authorize a $10,000 per day penalty against any person who attempts to write or distribute a pamphlet, news paper or book after being issued a Postal Service order to "cease and desist from such activity." - 2. Authorize the Postal Service to designate any or all of its 600,000 postal employees to "demand access" without a search warrant to any home, business, private library, file, bank vault or safes, to "inspect and copy books, records, documents or any object the Postal Service has reason to believe relates to any matter under investigation."

Our Postal Service would also be able to: - 3 Forbid the shipment or transformation of a banned book, not only by mail but by any other "instrumentality of interstate commerce," including air lines, parcel service, buses, trucks, or your own car! This clear and present danger to our freedoms of speech, press and thought has already been passed by the Senate and now is being discussed in the House. So this indicates that such organizations as THERA should have its members write to their Congressperson soon to have this Bill stopped Otherwise, we may find ourselves with a severe shortage of information!

K.L. Deer - Belfair, WA.

Ed: My deepest gratitude for the info on HR3973. I implore all of our readers to write their Congressperson about this Bill. This is a lot closer than you may think, as Gray Barker just sent his readers a special bulletin concerning a man from Kiss, who was stopped from publishing a pamphlet entitled, "Stale Food vs. Fresh Food." The FDA had a hand in this, and as the man was selling only an idea and not a product, his Constitutional Rights have been badly violated. If the government can stop this man from printing his ideas, we can all be stopped!

From Vol 4, NO. 2, Feb. 1983

Mary Dear,

Many thanks for the very early dispatch of Vol. 4 - No. 1 Hollow Hassle. It is so good of you and this is a particularly interesting edition. I was very intrigued with the article by Charles A. Marcoux "I Live With The Teros." It is so heartening to hear of the evolved races of people in the Inner Earth as previously we have read so much of the Deros and the negative types. It could be perhaps that Richard Shaver may have gone back to the Teros to get away from the Deros because he was a very knowledgeable man. The letter to you from M.L. Deer, Belfair, W.A. (HR3973) is a truly shocking indictment of the USA government and it carries out the prophesies we have been receiving for years that we are to become a race of slaves, mindless, dominated by the Rule of 12 (CFR). It seems it is not only the USA but new every world government is under the direction of what Wiley Crabb calls the Old Guard. I even had my English *Flying Saucer Review* stamped "Passed by Censor" here in New Zealand. It is still a wonder that *The Borderland Sciences Review* is so far able to escape the attention of the authorities or maybe the Guardians are helping that and other useful publications.

Recently I read "Alternative 3" which tells of the co-operation of the USA and the USSR in space research, operations on the Moon including mining, etc., and of course the inevitable work of the CIA in killing off everyone who learns a bit too much regarding such operations. FSR has just announced that France, which previously had a special department for the investigation and study of UFO reports has now joined the "Silence Brigade" and no further info will be coming from that direction.

Please dear friend and worker in the Light, carry en, and may Cosmic Blessings of Health, Happiness in your recent marriage, and continued protection be ever with you and yours. Happy Christmas and Good Luck for the Coming Year.

Phyllis Dixon Hall - New Zealand

Ed: Phyllis, I am glad to see that you are very aware of how much we are losing our freedoms everyday. Too many people imitate the ostrich nowadays, with their head in the sand and their ass in the air. It's not too early to say: Big Brother Is Watching!

Dear Mary,

I have gone into some caves in southwestern Wisconsin, northeastern Iowa and southeastern Minnesota. This was a few years ago. One cave in Minnesota called Niagara was rather deep and had more than one shaft that went straight down - very deep down. When a rock or stone was dropped in some of these you could hear water splash rather deep down - in others, nothing was heard. Parts of Niagara (beyond what they call "Grand Canyon1") I believe are still unexplored. It's too bad I don't have the time to explore such things myself. However, if you know of a spelunker or Hollow Earth researcher in Wisconsin, could you give me their address? I know the government is covering up what it knows about the inner earth, just as it is UFO's.

On the caves again, Niagara can be found on any good road map of Minn. It is south of Harmony on Route 139. Spook Cave is near McGregor, Iowa. Both have tours, etc. and are well known in the area. However, in checking with my relatives who went with me to those caves, I found that it is the third cave, located somewhere between the other two, that is the most interesting. We came across it much by accident and we cannot remember its name. It is off the beaten path, does not have a gift shop as such, but has an old one room school house near the entrance. In fact, you go out the back door down a dirt path to the entrance of the cave. There is someone there who does give tours. I'm sorry I can't be of more help one this cave.

Kenneth Van Hoof - Box 608, R. #2
Necadah, Wis.

Ed: Ken, I am printing your complete address so that any readers that may want to do some spelunking in your area can get in touch with you. I want to thank you for all the info you've sent and I would also like to inform our readers that you publish a periodical called Diamond Star, dealing with prophecies, hollow earth, and much, much more. Write Ken for details.

Dear Mary,

I have some information on the cave near Cushman, Ark. There's a young guy here, about 25 years old. He's a real husky, stocky guy, like a weight lifter. Anyhow, this guy is from Cushman, Ark. and he knows of Blowing Cave. He said he's been about one mile back in the cave and even lived in the cave for a period of about seven months. He also said that there's a hairy giant that lives in the cave. When I asked him if it was a Yeti or Sasquatch, he said, "no," but he may not be familiar with it called by that name, and nay call it something else. He said that a let of people who went into the cave never came out, and on weekends a lot of people go to the cave to drink and might try to pick a fight. From what he said, there are a lot of passages in the cave and it's not very well explored. He also mentioned that carbides and flashlights go out when you go so far back and breathing is harder the deeper you go. What kind of lighting shall we use? I've heard something about lanterns (sulfur). Have you heard of this?

LaBron Bynum - Springfield, Missouri

Ed: I haven't heard of the sulphur lanterns LaBron, but I will check on it and see what I can find out. As for Blowing Cave, it sounds more interesting everyday, and I for one can hardly wait to check it out. Keep the news coming...

Dear Mary,
It is impossible for me to continue as I have in the past and I'm just going to take a back seat and take things a day at a time. This hollow earth thing is so full of false claims. It is hard to distinguish fact from fiction. Four or five authors have cited the "hole at the pole" photographs that Ray Palmer published as proof of the hollow earth. I had several sets of these photos and they do not show any hole or any thing else! The so-called "hole" is simply a blank area because of the way these photos were taken, developed and printed. Ray Palmer borrowed my photos and published some of them in a later issue. It took me months and 3-4 letters before I got these photos back. (Ed: "You're lucky Frank. I never got the photos back that I sent him, and I even made long distance phone calls). I could never convince him that he was wrong, or at least he didn't admit it. Another mistake that has been erroneously perpetrated by many authors is that Raymond Bernard disappeared in Brazil. He died of pneumonia in 1965. His former secretary told me this. He also told me that Bernard didn't wander through the jungles seeking inner earth entrances. He got his information from others.
Another incident is my search for an original copy of the Hefferlin Manuscript. What Riley Crabb published was only part of

the complete manuscript. In the 30 years or so that I have been trying to track down a copy, I have been told all kinds of yarns. Some positively ridiculous! I have only known two people that claimed they had copies. Charles Marcoux and Riley Crabb borrowed the copy he used to put out "A Description Of Rainbow City From The Hefferlin Manuscript" around 1960. Charles Marcoux claimed that his copy was used to make this book. Riley Crabb says "No!" To this day I'm confused about this item. Some say it is a book and one guy said he could get me a copy for $2,000. Others say it was a manuscript on cheap paper, about 160 pages. Still others say it was merely a collection of newsletters that the Hefferlins put out over the years. As for the violinist mentioned, I'm sorry but my memory is getting so bad that I can't remember his name now and just about all my files are packed away. But I'm sure you know who I mean. I spent some years and I finally tracked this fellow down in N.Y.C. I wrote him asking about his part in this Rainbow City tale. Naturally he wrote and denied any knowledge of what I was talking about. Several years later I wrote again, but he had moved and left no forwarding address. Incidentally, the copy that Riley Crabb used was returned to Mrs. Flora Headen in Montana, I believe, and it was later ruined in a flood. I'm also wondering about Wilford A. South and the Lodge of the Lion.

I hear so many different opinions on this. I was corresponding with Mr. South 4-5 years and just before he died he was going to send me a manuscript on the Williams expedition. According to his landlady, most all of his inner earth material was not in his room after his death. She believes one of his relatives must have taken this stuff or else there just wasn't much hollow earth stuff left at the time of his death. (Ed: The latter may be closer to the truth Frank, as he sold a lot of his material). Perhaps he sold or gave it away. I did get some miscellaneous pages and quite a few drawings he made.

Best Wishes,
Frank Brownley Rochester, N.Y.

###

From Vol. 5, NO. 4, Aug. 1984

THE BEST OF THE HOLLOW HASSLE

Editor: The following letter was sent to Robert Schiller and he kindly passed it on to me, so all the readers could have the benefit of the information therein. It pertains to information on the Williams Expedition mentioned in the above letter and also to a crystal that Frank Williams found in a cave on one of his expeditions. Both of these items were sold to Rev. Wayne Taylor - City of the Sun Foundation - Columbus, N.M., by the widow of Frank Williams. Robert Schiller was good enough to write these people and inquire about the crystal, which was reported to have great powers. It is also said that the crystal has two small holes at the round end (it is shaped like a teardrop). Here is the reply that Robert received:

Dear Robert,

Yes, Rev. Wayne Taylor has a good sized crystal but I'm afraid it does not do for him what you had heard. It is beautiful and gives off strong emanations, as any crystal would of similar size. Wayne himself has never felt the vibrations, or seen "one picture" in it. Actually, it is up to the person holding or looking at it as to what is seen or felt! So, this is the same story as all others. WE RECEIVE WHAT WE ARE ABLE TO RECEIVE! A person or two looking in it has seen pictures of events of past, present or future, but these were people with a great deal of spiritual discernment, probably very advanced in this line. Each of us develops differently and are able to do different things. The Masters have said that none of us can really KNOW OURSELVES AS WE TRULY ARE! So, don't be discouraged. Your High Self probably is advanced in many ways you aren't fully aware of.
Bless you.

In Christ's Light, Mary Jane Switzler

Dear Mary,

On the property adjoining ours, there is a large cave. One young guy went back in it quite a ways, encountered some sort of obstacle and came out. A few months later a geologist from Missouri came down and warned them about the cave. He said he'd gone back in and some ways back had encountered what seemed to be a bottomless chasm. He said that several years ago two 12 year old boys had gone into the cave and never returned. Now the people who own the property have sealed up the cave for all intents and purposes. Also there is a strange indentation in the earth there known locally as "The Devil's Tea Cup." But it is all privately owned and they wouldn't want a bunch of people coming around. Really I don't blame them. I've heard a lot of strange stories about caves here in Arkansas. Only one person I've talked to has heard of the hollow earth theory. But the stories I've heard have had mainly to do with treasure, etc.

The number of caves here is incredible! We often hear strange rumblings and something like machinery, a kind of low roar. A lot of times our T.V. set picks up this noise. Don't know what it is.

Best Wishes,
Persephone - Little Joe, Arkansas

Dear Mary,
I am a new subscriber to Hollow Hassle, but I am not new to the inner earth idea. Long ago, I read about it in Blavatsky's SECRET DOCTRINE, Vol. 2, page 6: "The Sacred Imperishable Land is stated never to have shared the fate of our own continents. It is the cradle of the first man and the dwelling of the last divine mortal...the pole star has its watchful eye upon it..." Blavatsky considered knowledge of the inner earth to be one of the teachings of the Mystery Schools. Very secret and still is.

Although the Bible refers to the inner earth in several places as the "sides of the north" and the "ends of the earth", why does not Genesis tell us of the creation of the inner earth? Perhaps it does. In Genesis 1:12 on the third day plant life appears, grass herbs and trees. Notice this was before the sun, moon and stars, which came on the fourth day. Was there plant life without the sun? From where? From the interior sun, of course.

It is still talking about the inner paradise where the first man lived. On the second day a firmament was created with waters, oceans and seas above and below. This firmament, or support, is none other than terra firma, not the vaulted dome of the sky. Here in Genesis 1:6 is the creation of the hollow earth, waters above and below with the solid earth between, (for 2,000 miles). Here is described paradise or heaven with its herbs, trees, and grass. It is a tropical climate. There is no clothing worn and no farm work. Food is free for the taking. When the man and woman were driven out of this inner earth, the Lord placed at the end of the earth a flaming sword which turns everyway. It turns even today, the Aurora Borealis, to keep someone from entering the inner earth. The conditions cause mirror images called 'water sky' so that the opening is obscured. Government scientists took simultaneous pictures of both the north and south auroras. They matched perfectly. They decided the causes of both the auroras were the same and evidently earth caused. We can consider the auroras, the Eulerian Circle, the Van Allen Belts, the movements of continents, all indicating a hollow earth. We can consider these things along with the world scriptures and classic literature and poetry to tell us something of the secret that is hidden from us. While I am writing, I 11 just enclose my next renewal of $10.00.

Sincerely,
Hazel Coffey - Manteca, Calif.

Dear Mary,

I am editing a book on the legends of Mt, Shasta and have completed the main part of the ms. except for a few minor details, so I'm planning for a fall release of the book if all goes well. I think you'll find it quite interesting as it is a collection of several chapters each written by different individuals who have experienced the mountain first hand or have information on its mysteries. Some of the contributors to the book include, among others, Christine Hayes, Jim Moseley, Timothy Green Beckley and Peter Caddy (one of the original members of the Findhorn community who came to Mt. Shasta to lay the groundwork for a similar settlement at the base of the mountain).

I have some interesting info on ETIDORHPA to add to that mentioned by Eddie Rietz, who visited a spring-cave near Salem, KY and heard of an old Frenchman who lived in the area who would enter the cave and disappear for months at a time (see SHAVERTRON. Issue No. 18).

A friend of mine, who happens to be one of the most well-known spelunkers in the U.S., told me that he visited the Salem area several years ago and explored some of the caves there. Eddie Rietz did not give the name or location of the spring except for the fact that it was very near a "Hodge" Cave. This friend of mine informed me that Hodge Cave is now known as Shelby's Cave and is located a few miles south-east of Salem, Livingston Co., KY, and a few hundred feet to the east is a place called "Puckett Spring." I'm sure that this is the spring that Rietz visited in March of 1950, at which time he was told that some Masons tried to enter it by boat, but without success (the boat they used was still there when Rietz visited it).

If the entrance to Etidorhpa does exist Puckett Spring would most likely be it (even though I speculated in the past that it could be near the town of Lola, northwest of Salem). My spelunker friend

(M.D.) states that there is definitely a cave there, the entrance to which is a pit up the hill behind the spring. When he and his friends visited the pit they found that someone had covered it with logs, which they removed (the pit is next to a road fork behind the spring). They rappelled into the water-filled cavern which they explored with inner tubes, and they also explored a few other rooms but according to M.D. the cave didn't seem to "go," at least "without scuba gear."

I wonder whether they might have been more persistent if they knew about the "Etidorhpa" story? Well, if there is a most likely place to look for this entrance that would be the spot. One other thing he mentioned was that there is an underground river back in Hodge, or Shelby's cave, which he believed might connect with Puckett spring. Scuba gear would probably be needed to explore this river up and down stream to see if it does connect with Puckett, according to M.D.

Sincerely,
Bruce Walton
PO Box 1942
Provo, Utah 84603

VOL. 2 - #4   AUG. 1, 1981

THE BEST OF THE HOLLOW HASSLE

If you enjoyed this book, write for our free catalog of amazing books and mystifying DVDs:

Global Communications
P.O. Box 753
New Brunswick, NJ   08903

E-mail: mrufo8@hotmail.com

Visit our website:
www.conspiracyjournal.com